INCOMPLETE
A?

To John Boarder
"See mystery to
mathematics fly..."
— Alexander Pope

Michael Redhead

Incompleteness, Nonlocality, and Realism

A PROLEGOMENON TO THE PHILOSOPHY
OF QUANTUM MECHANICS

MICHAEL REDHEAD

CLARENDON PRESS · OXFORD

Oxford University Press, Walton Street, Oxford OX2 6DP
Oxford New York Toronto
Delhi Bombay Calcutta Madras Karachi
Petaling Jaya Singapore Hong Kong Tokyo
Nairobi Dar es Salaam Cape Town
Melbourne Auckland
and associated companies in
Berlin Ibadan

Oxford is a trade mark of Oxford University Press

Published in the United States
by Oxford University Press, New York

© Michael Redhead 1987

First published 1987
Reprinted (new as paperback) 1989
Reprinted 1990

All rights reserved. No part of this publication may be reproduced, stored in a retrieval system, or transmitted, in any form or by any means, electronic, mechanical, photocopying, recording, or otherwise, without the prior permission of Oxford University Press

This book is sold subject to the condition that it shall not, by way of trade or otherwise, be lent, re-sold, hired out or otherwise circulated without the publisher's prior consent in any form of binding or cover other than that in which it is published and without a similar condition including this condition being imposed on the subsequent purchaser

British Library Cataloguing in Publication Data
Redhead, Michael
Incompleteness, nonlocality and realism:
a prolegomenon to the philosophy of
quantum mechanics.
1. Quantum theory 2. Physics —
Philosophy
I. Title
530.1'2'01   QC174.13
ISBN 0–19–824937–3
ISBN 0–19–824238–7 (Pbk.)

Library of Congress Cataloging-in-Publication Data
Redhead, Michael.
Incompleteness, nonlocality, and realism.
Bibliography: p.
Includes index.
1. Quantum theory. 2. Physics—Philosophy.
I. Title.
QC174.12.R43   1987   530.1'2   86–23903
ISBN 0–19–824937–3
ISBN 0–19–824238–7 (Pbk.)

Printed in Great Britain
by Biddles Ltd,
Guildford and King's Lynn

Preface

THIS book is intended to be useful, to clear metaphysical fog, and to persuade the reader to join in and develop the argument. I am grateful to my students in London and Oxford who asked me to recommend a book on quantum mechanics to them, so that I ended up writing one myself. I hope physicists will read the book as well as philosophers. I have written it for both of them. I assume a nodding acquaintance with nonrelativistic quantum mechanics and the elements of linear algebra, but mostly everything is explained as I go along. I have definitely tried never to let technicalities obscure the arguments.

I acknowledge the help of a number of my friends and colleagues, especially Heinz Post, Jon Dorling, Arthur Fine, Abner Shimony, Richard Healey, John Lucas, Jeremy Butterfield, and Rom Harré. I value their comments to me on various occasions on matters discussed in the book. Thanks are due also to my research students, in particular Harvey Brown and Peter Heywood, who have collaborated with me in developing some of the ideas. The mistakes and misapprehensions are of course all mine. I would like to mention also Leonardo Castillejo, who showed me how to do theoretical physics, and to thank Marie Louise Varichon, who valiantly typed a scrawled manuscript.

The book is dedicated to my wife, Jennifer. Without her encouragement I would never have embarked upon it.

London, February 1986   MLGR

Preface to Second Impression

In this new impression, the opportunity has been taken to correct a number of misprints. I am grateful to those attentive readers who supplied me with detailed lists.

Concerning the main body of the text the most significant additional comment I would like to make relates to the discussion of violating the completeness condition for stochastic hidden-variable theories given in Section 4.4, and the subsequent investigation of the nonrobustness of the singlet state for two spin-$\frac{1}{2}$ particles, and the relation to Shimony's idea of passion-at-a-distance. In the text I considered two alternatives for accounting for the quantum-mechanical correlations. First, a *common* cause arising from the hidden state of the source and secondly, a *direct* stochastic link. The first alternative is ruled out by the failure of Jarrett completeness (or what Shimony prefers to call outcome independence). The second leads to the way of passion. But there is a third possibility, a common cause overlaid by an additional direct interaction. It has been shown in Redhead (1988) that this third possibility does not allow the common cause to be overlaid by a 'passion' link, but only by a stochastic *causal* link. The way of passion and the resulting possibility of peaceful coexistence with the constraints of special relativity is only possible if we give up any contribution from the hidden state of the source, and interpret the correlations as produced solely by a direct 'interaction' in the way described mathematically in the text. But this means effectively giving up the stochastic hidden-variable framework as a means of understanding the correlations and at the same time avoiding conflict with special relativity. The way of passion is not a variant of the stochastic hidden-variable approach, but is properly to be understood as a rejection of it (even in the nonfactorizable form that could accommodate mirror-image correlations and the violation of the Bell inequality).

Furthermore, in order to avoid misunderstanding, we should remark that the robustness condition for stochastic causality has nothing to do with stability, that small changes in the cause produce small changes in the effect. Robustness is concerned only with the invariance of the functional form of the connection under sufficiently small perturbations. Moreover it should be noted that the robustness condition is not necessarily violated in the case of 'non-Markovian' causal processes, where the manner in which one event is brought about gets incorporated in the cause of another event. The claim is that *at some stage* in the process of incorporating antecedents in the total cause, robustness must be rescued. Otherwise we would live in a 'marshmallow' world where the notion of cause would not, I believe, be appropriate. In addition it is shown in my (1988) that, in general, in an indeterministic framework, robustness is logically independent of signalling capability. More precisely, nonrobustness under a given class of sufficiently weak perturbations cannot in general be deduced from the impossibility of signalling using that class of perturbations. However, it remains true that no-signalling under perturbations which allow sufficiently arbitrary control of local marginal probabilities does imply nonrobustness.

Apart from this more substantial clarification, I have taken the opportunity of adding a few additional references, in order to bring the discussion of matters more thoroughly up to date. The new references are listed as a supplement to the main bibliography.

Wolfson College, Cambridge   MLGR
November 1988

I am grateful to the following sources for copyright permission to include material from the relevant publications, in this book:

Indiana University Mathematics Journal: Kochen and Specker (1967)
D. Reidel Publishing Company: Redhead (1983).
Plenum Publishing Corporation: Heywood and Redhead (1983).
The New York Academy of Science: Redhead (1986).

Contents

Introduction   1

1. The Formalism of Quantum Mechanics   5
   1.1. The Dirac Formulation   6
   1.2. The von Neumann Formulation   12
   1.3. Functions of Observables   16
   1.4. The Logico-Algebraic Approach   22
   1.5. Gleason's Theorem   27
   1.6. The Quantum Mechanics of Many-Particle Systems   30
   1.7. Angular Momentum in Quantum Mechanics   32
   1.8. Angular Momentum of a Composite System   40
       Notes and References   42

2. The Interpretation of Quantum Mechanics   44
   2.1. View A: Hidden Variables   45
   2.2. View B: Propensities and Potentialities   48
   2.3. View C: Complementarity   49
   2.4. Measurement and State Preparation   51
   2.5. The Uncertainty Relations   59
       Notes and References   69

3. The Einstein–Podolsky–Rosen Incompleteness Argument   71
       Notes and References   81

4. Nonlocality and the Bell Inequality   82
   4.1. The Bell Inequality   82
   4.2. Counterfactuals and Indeterminism   90
   4.3. Alternative Forms of the Bell Inequality   96
   4.4. Stochastic Hidden-Variable Theories   98
   4.5. Experimental Tests of the Bell Inequality   107
   4.6. Statistical Nonlocality   113

4.7. Summary of Conclusions		116
Notes and References		117

5. The Kochen–Specker Paradox — 119
    - 5.1. Demonstration of the Contradiction — 121
    - 5.2. The Justification of FUNC — 131
    - Notes and References — 138

6. Nonlocality and the Kochen–Specker Paradox — 139
    - 6.1. Contextuality and Nonlocality — 139
    - 6.2. The Comeasurable Value Rule — 142
    - 6.3. The Incompatibility of CVR and Locality — 146
    - 6.4. Implications — 150
    - Notes and References — 151

7. Realism and Quantum Logic — 153
    - 7.1. The Revisability of Logic — 153
    - 7.2. Classical Propositional Logic — 155
    - 7.3. The Logic of Classical Physics — 157
    - 7.4. Quantum Propositional Logic — 160
    - 7.5. Putnam States and Realism — 164
    - Notes and References — 167

8. Envoi — 168
    - Notes and References — 169

Mathematical Appendix — 170

Bibliography — 179

Index — 187

Introduction

THE object of this book is to present in as straightforward and uncluttered a manner as possible some of the exciting work that has been done in the philosophy of quantum mechanics during the past few years, in particular since the discovery of the now famous Bell inequality in 1964, and the apparently unrelated but equally important work of Kochen and Specker in 1967. A great deal has been written since, developing and applying the techniques of these two seminal papers; but now, some twenty years later, the dust is beginning to settle, and we can try to take stock of just what has been learned about quantum mechanics, and how it relates to the classic era of debate and discussion between Einstein and Bohr which culminated in 1935 with the Einstein–Podolsky–Rosen incompleteness argument and its putative rebuttal by Bohr. Bohr was in effect declared the winner in the controversy, and for nearly thirty years Copenhagen orthodoxy reigned supreme. Another factor here was the dead hand of von Neumann's proof of the impossibility of introducing a more complete specification of the state of a system than that provided by quantum mechanics—the impossibility of so-called 'hidden variables' in quantum mechanics. This proof had been spelled out in von Neumann's magisterial work *Mathematische Grundlagen der Quantenmechanik*, published in 1932, a book more frequently referred to than actually read by physicists because of its mathematical sophistication. The von Neumann proof was carefully examined, and its defects exposed, in a famous article published by Bell in 1966. Bell's work also showed in detail how the hidden-variable theories of Bohm produced in the early 1950s actually worked, and just how they circumvented the no-hidden-variable theorem. Bell moreover pointed out the connection with Gleason's theorem, the major technical development in the foundations of quantum mechanics since von Neumann's work. These new developments in the 1960s revived interest in the old debates about 'realist' interpretations of quantum mechanics, and showed much more clearly just what were the difficulties in implementing the Einstein programme of a 'complete' version of quantum mechanics. Essentially what has been achieved is not a resolution of the difficulties surrounding the interpretation of quantum mechanics, but a clarification of just what those difficulties are.

In Chapter 1, I gather together all the material needed in respect of the formalism of quantum mechanics. In doing the philosophy of physics, two sorts of problems commonly arise. In the first place there are very general metatheorems that involve a highly technical approach; for example, attempts to show the measurement problem in quantum mechanics to be insoluble require a very general and abstract discussion of what counts as a measurement. On the other hand, there are very simple and highly idealized thought experiments designed to point up some particular problem about the interpretation of a physical theory. Here one looks for the simplest physics that will do the job—there is no need to aim at generality or sophistication. In the main we shall be concerned with the second sort of example, and this is reflected in the fairly meagre mathematical and physics background we shall require. In particular, we shall not explicitly bother about problems associated with operators having a continuous spectrum, and indeed we shall largely confine ourselves to finite-dimensional Hilbert spaces, where the spectral theory for self-adjoint operators is straightforward and easily explained. A number of different approaches and notations are discussed, since an ability to translate quickly between different formulations of the basic ideas is essential to following the literature in the philosophy of quantum mechanics. The paradigm example of a finite-dimensional Hilbert space is that employed in describing the spin observables in quantum mechanics. We develop sufficient material here for treating spin-$\frac{1}{2}$ and spin-1 systems, which we shall frequently use as examples in the later chapters.

Chapter 2 is concerned with possible interpretations of the formalism presented in Chapter 1. A number of different approaches are distinguished, and brief discussion given of hidden variables, propensities, complementarity, the problem of measurement and the interpretation of mixed states, and the significance of the uncertainty relations in quantum mechanics. All these matters are of course of great importance in understanding the significance of the quantum-mechanical formalism; but they are not the central topics treated in this book, and consequently no attempt is made to deal comprehensively with them.

In Chapter 3 the main discussion of the book really begins. We deal here with the Einstein–Podolsky–Rosen argument for the incompleteness of the formalism of quantum mechanics. Effectively a dilemma is posed: either we accept a form of nonlocal action or we must admit incompleteness. Rejecting nonlocality, Einstein, Podolsky, and Rosen grasped the incompleteness horn. But this suggests a programme of 'completing' quantum mechanics by some more comprehensive specification of the state of a system than that entertained in the quantum-mechanical formalism.

In Chapter 4 we show how this 'completion' programme runs into a nonlocality difficulty of its own via the experimentally well-confirmed violation of the so-called Bell inequality. So both horns of the original dilemma appear to lead to some form of nonlocality. But this conclusion needs careful discussion in order to clarify what is really going on. An attempt is made in this chapter to give as clear an evaluation as possible of the nonlocality issue.

But nonlocality is not the only difficulty confronting a 'completed' quantum mechanics. A quite different and purely algebraic problem is posed by the so-called Kochen–Specker paradox, which is discussed in Chapter 5. The question of how to evade the paradox and at the same time retain a realist approach to the interpretation of quantum mechanics is discussed.

Then, in Chapter 6, the connection between the nonlocality issue and the Kochen–Specker paradox is examined, and a very intimate relation is brought to light, which enables a demonstration of nonlocality to be achieved which is almost, but not quite, purely algebraic in character.

In Chapter 7 a quite different line of attack on the realism issue is investigated. This is the quantum logic approach, which has been the subject of a great deal of discussion during the past fifteen years or so. The position adopted here is frankly critical of the exaggerated claims of some quantum logicians. But again, the object of the discussion is clarification of the very important issues involved, which bear as much on the philosophy of logic as on the philosophy of quantum mechanics.

In a short final chapter, I attempt to draw together the threads of the argument, and to summarize what can reasonably be learned about the interpretation of quantum mechanics from the extensive, confusing, and sometimes contradictory literature that has emerged since the mid-1960s, and which serves in many cases to baffle rather than enlighten the intending student of the philosophy of quantum mechanics.

At the end of each chapter I include a section entitled 'Notes and References'. This is really an annotated (and fairly selective) bibliography, but one which helps to trace the historical development of most of the ideas discussed. The references cited in these sections by author and year (for example, 'Bell (1966)') are collected alphabetically in more scholarly detail at the end of the book.

In a Mathematical Appendix, the salient results about set theory, vector spaces, and lattices are summarized, so as not to interrupt the main text with discussions of purely mathematical matters.

A final note concerns the numbering of equations, and cross-referencing. In each chapter equations are numbered serially—1, 2, 3, etc. Reference in a given chapter to an equation in the same chapter is given by citing just this number. References to equations in a different chapter include the chapter number. Thus, a reference to Eq. (2.33) in Chapter 3 would refer to Eq. (33) in Chapter 2.

1
The Formalism of Quantum Mechanics

THE formalism of quantum mechanics (QM) is designed to accommodate two features of atomic and subatomic systems. First, the possible results of measuring certain physical magnitudes on such systems are confined to a restricted set of possible values (real numbers). Secondly, it is in general not possible to predict, for any physical magnitude, what value will turn up on measurement, only the probability that any particular value from among the set of possible values will turn up. Physical magnitudes that can be measured are called observables, and the specification of the probabilities of measurement results for observables depends on assigning to the system in question a state: in other words, the state of the system is just an expression of the various probabilities, for all the observables, of the possible outcomes of measurement. The mathematical scheme for QM consists then in setting up a mathematical structure such that certain elements in that structure are associated with the states of the physical system and certain other elements are associated with the observables. Certain algorithms are then proposed which serve to answer our two basic questions:

1. What values are possible measurement results for any given observable?
2. For any given state and any given observable, what is the probability that one of the possible measurement results will actually turn up when a measurement is performed?

We shall refer to the algorithm which answers the first question as the *quantization algorithm*. The algorithm which answers the second question we shall refer to as the *statistical algorithm*.

The reader may wonder why we introduce two algorithms. The numbers generated by the quantization algorithm are usually identified with those which turn up in some state with non-vanishing probability according to the statistical algorithm. Or, to put it another way, if the probability of a certain measurement result is always zero, this value cannot be the result of a measurement. But that is just wrong. Zero probability is quite different from impossibility. It is consistent, for example, with any finite number of occurrences in an infinite collective of outcomes, if we adopt the usual relative frequency interpretation of probability. The converse proposition, that if certain measurement outcomes never occur, then the probability for these outcomes is zero, is of course correct, and the two algorithms must mesh in such a way that that result is satisfied.

There are a number of different ways of expressing our two basic algorithms. We shall begin with that due to Dirac.

1.1. The Dirac Formulation

The mathematical structure is an abstract *vector space* V, defined over the field of complex numbers, equipped with an inner product (for details see the Mathematical Appendix) which we write in the following way:

If α and β are any two vectors in V then the inner product of α and β is a complex number written as

$$\langle \alpha | \beta \rangle \tag{1}$$

and we have

$$\langle \beta | \alpha \rangle = (\langle \alpha | \beta \rangle)^* \tag{2}$$

where * denotes the complex conjugate.

Dirac himself writes vectors such as α and β with the notation $|\alpha\rangle$, $|\beta\rangle$. These he refers to as *kets*. He then introduces dual vectors which he calls *bras*, written $\langle\alpha|$, $\langle\beta|$ etc. Dual vectors are linear functionals on a space of vectors. The value of $\langle\alpha|$ at the vector $|\beta\rangle$, often referred to as the contraction of $\langle\alpha|$ with $|\beta\rangle$, should then be written in function notation as $\langle\alpha|(|\beta\rangle)$. Dirac abbreviates this as $\langle\alpha|\beta\rangle$. Now, in general it can be shown that for any dual vector $\langle\alpha|$, there exists a ket $|\alpha\rangle$ such that $\langle\alpha|(|\beta\rangle) =$ inner product of $|\alpha\rangle$ and $|\beta\rangle$ $= \langle\alpha|\beta\rangle$ in the notation we introduced in (1). Hence we can understand the symbol $\langle\alpha|\beta\rangle$ either as an inner product of the two kets $|\alpha\rangle$, $|\beta\rangle$ or as the contraction of the bra $\langle\alpha|$ with the ket $|\beta\rangle$.

Let us first restrict ourselves to finite-dimensional V.

States are associated with unit vectors in V, and observables are associated with self-adjoint linear operators on V. Note that, since we

The Formalism of Quantum Mechanics 7

are dealing with the field of complex numbers, states are only determined up to an arbitrary phase factor of unit modulus. In fact, as we shall see later, it is the *rays*, the one-dimensional subspaces generated by the unit vectors, that are of crucial importance here.

We also assume, for the time being, that every self-adjoint operator is associated with a unique observable. This assumption will be relaxed later (see Chapter 5).

Consider a self-adjoint operator $\hat{Q}$. It possesses eigenvectors $|q_i\rangle$ satisfying

$$\hat{Q}|q_i\rangle = q_i|q_i\rangle \tag{3}$$

where q_i is some real number known as an eigenvalue of $\hat{Q}$, and i runs from 1 to N, where N is the dimension of the vector space. The q_i may be all distinct, in which case $\hat{Q}$ is said to be nondegenerate or maximal. If two or more of the q_i are equal in value, we speak of degeneracy. We shall interpret $|q_i\rangle$ as the ith eigenvector, having the eigenvalue q_i. It is thus labelled by the index i, not the numerical value of q_i. The $\{|q_i\rangle\}$ can be chosen so as to provide a complete orthonormal set of vectors in V. This means that

$$\langle q_i|q_j\rangle = \delta_{ij} \tag{4}$$

and for any vector $|\psi\rangle$ in V we can write $|\psi\rangle$ as some linear combination of the $|q_i\rangle$. Thus

$$|\psi\rangle = \sum_{i=1}^{N} c_i|q_i\rangle \tag{5}$$

where the complex coefficients c_i are given by

$$c_i = \langle q_i|\psi\rangle \tag{6}$$

We are now in a position to state the two algorithms.

Quantization Algorithm

The possible measurement results on Q are the eigenvalues of the associated operator $\hat{Q}$.

If we need specifically to distinguish operators from observables we shall use the 'hat' to denote the operator. This will be of particular importance in the developments discussed in Chapters 5 and 6. Otherwise we shall employ the same symbol Q for the observable and the associated operator.

Statistical Algorithm

The probability that Q will yield a measurement result q_i when the state of the system is $|\psi\rangle$ is given by

$$\text{Prob}\,(q_i)_Q^{|\psi\rangle} = \sum_{j|q_j = q_i} |c_j|^2 = \sum_{j|q_j = q_i} |\langle q_j|\psi\rangle|^2 \quad (7)$$

where the symbol $(q_i)_Q^{|\psi\rangle}$ signifies the proposition that the result of measuring Q in the state $|\psi\rangle$ is q_i and we use the notation $\sum_{j|q_j = q_i}$ to denote summation over all those values of j for which $q_j = q_i$.

It should be noted that, in the case of degeneracy, the choice of $\{|q_i\rangle\}$ is by no means unique, but in fact different choices all lead to the same result for the probability.

The mean or expectation value of these measurement results on Q in the state $|\psi\rangle$ is given by

$$\langle Q \rangle_{|\psi\rangle} = \sum_{i=1}^{N} q_i |c_i|^2$$

$$= \langle \psi|Q|\psi\rangle \quad (8)$$

We can also ask what is the probability of a measurement result lying in the set of values Δ where $\Delta = \{q_\alpha, q_\beta \ldots\}$ is a subset of S_Q, the set of distinct eigenvalues or *spectrum* of Q.

The answer is clearly

$$\text{Prob}\,(\Delta)_Q^{|\psi\rangle} = \sum_{j|q_j = q_\alpha, q_\beta \ldots} |\langle q_j|\psi\rangle|^2 \quad (9)$$

where $(\Delta)_Q^{|\psi\rangle}$ is a convenient symbol to denote the proposition that the result of measuring Q in the state $|\psi\rangle$ is a member of the set Δ. Strictly speaking this is not quite consistent with our use of the symbol $(q_i)_Q^{|\psi\rangle}$, but the distinction between a singleton set and its sole member should always be clear from the context.

Finally, let us suppose we are ignorant of which state among a set of states $\{|\psi_k\rangle\}\, k = 1, 2 \ldots m$, the system is actually in, and express our ignorance by introducing probabilistic weights w_k such that w_k is the probability of finding the quantum-mechanical state of the system to be $|\psi_k\rangle$, so $\sum_{k=1}^{m} w_k = 1$, then the probability that a measurement result on Q lies in the set Δ is given by the more general formula

$$\sum_{k=1}^{m} w_k \sum_{j|q_i = q_\alpha, q_\beta, \ldots} |\langle q_j | \psi_k \rangle|^2 \tag{10}$$

In order to complete the formal mathematical scheme of QM there are two additional questions to be dealt with:

1. How is the operator-observable correspondence to be set up?
2. How does the state of a system vary with time?

The first question is answered by means of a correspondence principle which links the Poisson bracket of two dynamical quantities in classical mechanics with the commutator bracket of the associated operators. Schematically

$$(\text{Poisson bracket}) \rightarrow \frac{1}{i\hbar} \times (\text{commutator bracket})$$

where $\hbar = h/2\pi$; and h is the famous Planck's constant with the approximate numerical value $6 \cdot 6 \times 10^{-34}$ Joule-seconds.

As an example consider the Poisson bracket relations for the Cartesian components J_x, J_y, J_z of angular momentum which satisfy

$$\{J_x, J_y\} = J_z \tag{11}$$

where the Poisson bracket $\{A, B\}$ of two dynamical quantities A, B is defined by

$$\{A, B\} = \sum_i \left(\frac{\partial A}{\partial q_i} \frac{\partial B}{\partial p_i} - \frac{\partial B}{\partial q_i} \frac{\partial A}{\partial p_i} \right) \tag{12}$$

in terms of the generalized coordinates q_i and conjugate momenta p_i specifying the state of the dynamical system. Substituting Poisson brackets by commutator brackets, this becomes

$$[J_x, J_y] = i\hbar J_z, \text{ etc.} \tag{13}$$

Where the commutator bracket is defined by

$$[A, B] = AB - BA \tag{14}$$

for arbitrary operators A and B.

Notice that these algebraic relations between operators, together with the eigenvalue spectrum of an operator and the probabilities specified by the statistical algorithm, are all unchanged by a unitary transformation of the vector space.

If we represent the abstract mathematical structure we have outlined by concrete realizations, in terms of $N \times N$ Hermitean matrices in the case of self-adjoint operators and $1 \times N$ column vectors in the case of state-vectors, different realizations correspond to different choices of the complete set of orthonormal basis vectors for the space, different bases being connected by unitary transformations.

Working in terms of concrete matrix realizations takes us from the Dirac abstract vector space formulation of QM to what is often called 'matrix mechanics'. Indeed, historically speaking, Heisenberg developed matrix mechanics in advance of the more abstract Dirac approach.

So far we have restricted the discussion to finite-dimensional state spaces. But Dirac attempted to extend his treatment to the infinite-dimensional case. Heuristically he considered operators with a continuous eigenvalue spectrum, and replaced the summation in Eq. (5) by an integration. For example, in the case of position in a one-dimensional configuration space, Dirac writes the position observable as an operator X with continuous eigenvalues x satisfying

$$X|x\rangle = x|x\rangle \tag{15}$$

An arbitrary state $|\psi\rangle$ is now expanded as

$$|\psi\rangle = \int_{-\infty}^{\infty} \langle x|\psi\rangle |x\rangle \, dx \tag{16}$$

The function

$$\psi(x) \underset{Df}{=} \langle x|\psi\rangle \tag{17}$$

is known as a wave function.

The kets $|x\rangle$, $|x'\rangle$ satisfy the orthogonality relation

$$\langle x|x'\rangle = \delta(x - x') \tag{18}$$

where $\delta(x)$ is the Dirac delta function which satisfies

$$\int_{-a}^{b} f(x)\delta(x) \, dx = f(0) \tag{19}$$

for an arbitrary function $f(x)$, and arbitrary limits $-a$ and b ($a, b > 0$). Notice that the norm of $|x\rangle$ is infinite since from (18), $\langle x|x\rangle = \delta(0)$

$= \infty$. The delta function is mathematical nonsense considered as a function (as opposed to a functional). It is however very convenient as a heuristic device, and we shall employ it freely in this work when discussing position and momentum operators. For the momentum P canonically conjugate to X we have the wave function,

$$\langle x|p \rangle = \frac{1}{\sqrt{2\pi}} e^{ipx} \tag{20}$$

where $|p\rangle$ is the eigenstate of momentum belonging to the eigenvalue p. (We are here using units with $\hbar = 1$). The kets $|p\rangle$ and $|p'\rangle$ satisfy a similar orthogonality relation to (18)

$$\langle p|p' \rangle = \delta(p - p') \tag{21}$$

We note the useful formal representation

$$\delta(x) = \frac{1}{2\pi} \int_{-\infty}^{+\infty} e^{ipx} \, dp \tag{22}$$

Now
$$\psi(p) \underset{Df}{=} \langle p|\psi \rangle = \int_{-\infty}^{\infty} \langle p|x \rangle \langle x|\psi \rangle \, dx$$

$$= \frac{1}{\sqrt{2\pi}} \int_{-\infty}^{\infty} e^{-ipx} \psi(x) \, dx \tag{23}$$

is the wave function in momentum space.

$\psi(x)$ and $\psi(p)$ have the interpretation that $|\psi(x)|^2$ and $|\psi(p)|^2$ measure the probability densities for finding values of X in the neighbourhood of x and of P in the neighbourhood of p—that is to say, the probability of finding a value for X in the range x to $x + dx$ is $|\psi(x)|^2 \, dx$, for example.

The eigenvalue problem for the energy of a system is now formally reduced to the solution of a certain differential equation, the so-called time-independent Schrödinger equation. Unfortunately the mathematical steps employed by Dirac in this extension to the infinite-dimensional case are quite incoherent, and we shall see in a moment how a much more satisfactory way of dealing with the problem was devised by von Neumann.

But we still have the unfinished business of how states evolve in time. This is described in the Dirac scheme by a time-dependent unitary transformation applied to the state vector

$$|\psi(t)\rangle = U(t, t')|\psi(t')\rangle \tag{24}$$

Formally

$$U(t, t') = e^{-i/\hbar \cdot H \cdot (t - t')} \tag{25}$$

for the case of a system governed by a time-independent Hamiltonian H.

We mention in passing a rather confusing terminology which has grown up in discussing time evolution in QM. The method we have just described of changing state vectors and fixed observables is called the *Schrödinger picture*, since in the case of wave mechanics it is associated with the so-called *time-dependent Schrödinger* equation (to be sharply distinguished from the time-independent equation, which specifies the energy eigenvalues of a system). But it is possible to describe time evolution in an alternative way by introducing a time-dependent unitary transformation affecting both states (vectors) and observables (operators) in such a way as to kill the time dependence of the state vectors. The net result of this manoeuvre is to transfer time dependence from states to observables. The resulting description of time evolution in QM is known as the *Heisenberg picture*. Intermediate situations in which time dependence is partly associated with states and partly with observables are also possible. One such intermediate picture is of importance in scattering theory and is, confusingly enough, referred to as the *Dirac picture*. Here the states remain constant in the absence of interaction, while even in the presence of interaction the observables evolve in time according to the unperturbed Hamiltonian. We shall have occasion to use the Dirac picture in the proof of the general no-signalling theorem in Section 4.6.

1.2. The von Neumann Formulation

Again, we shall start with the perfectly consistent Dirac scheme in a finite dimensional state space of dimension N. Observables are again associated with self-adjoint operators, and the quantization algorithm identifies possible measurement results with the (real) eigenvalues of these operators. The difference from Dirac comes in the way we express the statistical algorithm.

We begin by introducing the notion of a projection operator. For the observable Q consider the linear operator

$$P_{|q_i\rangle} = |q_i\rangle\langle q_i| \qquad (26)$$

That this object is a linear operator is clear when we consider the action of $P_{|q_i\rangle}$ on an arbitrary state $|\psi\rangle$

$$\begin{aligned}P_{|q_i\rangle}|\psi\rangle &= |q_i\rangle\langle q_i|\psi\rangle \\ &= c_i|q_i\rangle\end{aligned} \qquad (27)$$

i.e. $P_{|q_i\rangle}$ acting on $|\psi\rangle$ projects out the component $c_i|q_i\rangle$ in the expansion according to Eq. (5).

This is why we call $P_{|q_i\rangle}$ a projection operator. Its domain is the whole vector space V. Its range, however, is just the one-dimensional subspace spanned by $|q_i\rangle$.

Clearly

$$\sum_i P_{|q_i\rangle} = I \qquad (28)$$

where I denotes the identity operator, since $\sum_i P_{|q_i\rangle}|\psi\rangle = \sum_i c_i|q_i\rangle = |\psi\rangle$. So $\sum_i P_{|q_i\rangle}$ acting on an arbitrary vector $|\psi\rangle$ just reproduces $|\psi\rangle$.

Now we can write

$$\begin{aligned}Q = Q \cdot I = Q\sum_i P_{|q_i\rangle} &= \sum_i Q P_{|q_i\rangle} \\ &= \sum_i Q|q_i\rangle\langle q_i| \\ &= \sum_i q_i|q_i\rangle\langle q_i| \\ &= \sum_i q_i P_{|q_i\rangle}\end{aligned} \qquad (29)$$

This we refer to as the *spectral theorem*; and it is this result that does generalize rigorously to the infinite-dimensional case, although this requires a definition of the spectrum of an operator which does not employ the notion of eigenvalue, and also discussion of how to replace the sum by an integral in Eq. (29). We shall not need to go into these technicalities, but refer the reader to von Neumann (1932) for the details.

Another way of deriving the spectral theorem is to work in terms of the representation produced by taking the $|q_i\rangle$ as basis. In such a basis

Q is a diagonal matrix.

$$Q = \begin{pmatrix} q_1 & & & \\ & q_2 & & \\ & & q_3 & \\ & & & \ddots \end{pmatrix}$$

and

$$P_{|q_1\rangle} = \begin{pmatrix} 1 & & & \\ & 0 & & \\ & & 0 & \\ & & & \ddots \end{pmatrix}$$

$$P_{|q_2\rangle} = \begin{pmatrix} 0 & & & \\ & 1 & & \\ & & 0 & \\ & & & \ddots \end{pmatrix}$$

etc.

The fact that

$$Q = q_1 P_{|q_1\rangle} + q_2 P_{|q_2\rangle} + \cdots$$

is now trivially obvious. Since algebraic relations between operators do not depend on the choice of representation, the fact that we have here proved the spectral theorem relative to a particular (convenient) basis does not affect the correctness of the result as applied to the abstract operators.

To return to the statistical algorithm, we consider first the case of a nondegenerate eigenvalue q_i and write

$$\begin{aligned}
\text{Prob}\,(q_i)_Q^{|\psi\rangle} &= |\langle q_i|\psi\rangle|^2 \\
&= \langle q_i|\psi\rangle\langle\psi|q_i\rangle \\
&= \sum_j \langle q_i|\psi\rangle\langle\psi|P_{|q_j\rangle}|q_i\rangle \\
&= \sum_j \langle q_i|\psi\rangle\langle\psi|q_j\rangle\langle q_j|q_i\rangle \\
&= \sum_j \langle q_j|q_i\rangle\langle q_i|\psi\rangle\langle\psi|q_j\rangle \\
&= \text{Tr}\,(|q_i\rangle\langle q_i|)(|\psi\rangle\langle\psi|) \\
&= \text{Tr}\,(P_{|q_i\rangle} \cdot P_{|\psi\rangle}) \qquad (30)
\end{aligned}$$

where $P_{|\psi\rangle}$ is the projection operator whose range is the one-dimensional subspace spanned by $|\psi\rangle$, and Tr denotes the trace of a matrix, i.e. the sum of the diagonal elements.

This is our basic result, which can now be generalized in a number of significant directions.

Suppose q_i is degenerate, then define

$$P_Q(q_i) = \sum_{j|q_j=q_i} P_{|q_j\rangle} \tag{31}$$

$P_Q(q_i)$ is clearly also a projection operator, whose range is the subspace spanned by the eigenvectors $|q_j\rangle$ with $q_j = q_i$, and in this case, from the linearity of the trace operation,

$$\text{Prob}\,(q_i)_Q^{|\psi\rangle} = \text{Tr}\,(P_Q(q_i) \cdot P_{|\psi\rangle}) \tag{32}$$

Similarly, the probability of a measurement result lying in the set $\Delta = \{q_\alpha, q_\beta \ldots\}$ is

$$\text{Prob}\,(\Delta)_Q^{|\psi\rangle} = \text{Tr}\,(P_Q(\Delta) \cdot P_{|\psi\rangle}) \tag{33}$$

where,

$$P_Q(\Delta) = \sum_{j|q_j=q_\alpha,\,q_\beta\ldots} P_{|q_j\rangle} \tag{34}$$

is a projection operator whose range is the subspace spanned by all the $|q_j\rangle$ with $q_j = q_\alpha, q_\beta$, etc.

In the more general case, where we assign probability weights w_k to a set of states $\{|\psi_k\rangle\}$ in which a system might be found, the probability of a measurement result on Q lying in the set Δ is just $\text{Tr}\,(P_Q(\Delta) \cdot W)$ where

$$W = \sum_k w_k P_{|\psi_k\rangle} \tag{35}$$

W is known as the *statistical operator* for the system. Notice that W is not itself in general a projection operator. If W reduces to a single term, we talk of a *pure state*. The more general case with several w_k's unequal to zero is known as a *mixed state*.

Our basic result is now

$$\text{Prob}\,(\Delta)_Q^W = \text{Tr}\,(P_Q(\Delta) \cdot W) \tag{36}$$

where W specifies the state of the system via the statistical operator. So everything is now expressed in terms of projection operators, rather than in terms of vectors, as in the Dirac scheme.

Notice further the result for the expectation value of Q in the state W

$$\langle Q \rangle_W = \sum_i q_i \, \text{Tr}\,(P_{|q_i\rangle} \cdot W)$$

$$= \text{Tr}\left(\sum_i q_i P_{|q_i\rangle} \cdot W\right)$$

$$= \text{Tr}\,(Q \cdot W) \qquad (37)$$

1.3. Functions of Observables

Consider again the spectral expansion of a self-adjoint operator

$$Q = \sum_i q_i P_{|q_i\rangle}$$

From the orthogonality properties of the projection operators

$$P_{|q_i\rangle} P_{|q_j\rangle} = |q_i\rangle\langle q_i | q_j\rangle\langle q_j|$$

$$= \delta_{ij} |q_i\rangle\langle q_j|$$

$$= \delta_{ij} |q_i\rangle\langle q_i|$$

$$= \delta_{ij} P_{|q_i\rangle} \qquad (38)$$

we obtain immediately simple results like

$$Q^2 = \sum_i q_i^2 P_{|q_i\rangle}, \quad Q^3 = \sum_i q_i^3 P_{|q_i\rangle}, \text{ etc.}$$

and more generally, if p is any polynomial function, then

$$p(Q) = \sum_i p(q_i) P_{|q_i\rangle}.$$

We now generalize this result to give a definition of what we mean by $f(Q)$ for an arbitrary function $f: \mathbb{R} \to \mathbb{R}$, where $\mathbb{R}$ denotes the real line

$$f(Q) =_{Df} \sum_i f(q_i) P_{|q_i\rangle} \qquad (39)$$

The associated observable will also be denoted by $f(Q)$. Clearly in this definition we need only consider the *restriction* of f to the set S_Q

comprising the spectrum of Q, but it is often convenient to consider functions defined over the whole of the real line.

As an example let us express one of the projection operators $P_Q(q_i)$ as a function of Q. Clearly

$$P_Q(q_i) = \chi_{q_i}(Q)$$

where the characteristic function $\chi_{q_i}: \mathbb{R} \to \{0, 1\}$ is defined by

$$\chi_{q_i}(x) = 1 \quad \text{for } x = q_i$$
$$= 0 \quad \text{for } x \neq q_i \qquad (41)$$

But all we really need is the restriction $\chi_{q_i}|S_Q$ specified by

$$\chi_{q_i}(q_j) = \delta_{q_i q_j} \qquad (42)$$

Any function defined on $\mathbb{R}$ which takes on the same values for the points of S_Q as χ_{q_i} will serve equally as well as χ_{q_i} itself in defining $\chi_{q_i}(Q)$. Thus, assuming q_i is nondegenerate, the function $F_{q_i}: \mathbb{R} \to \mathbb{R}$ defined by

$$F_{q_i}(x) = \frac{(x - q_1)(x - q_2) \ldots (x - q_{i-1})(x - q_{i+1}) \ldots (x - q_N)}{(q_i - q_1)(q_i - q_2) \ldots (q_i - q_{i-1})(q_i - q_{i+1}) \ldots (q_i - q_N)}$$

(43)

has the same restriction to S_Q as χ_{q_i} and may be used to 'represent' χ_{q_i} in so far as we are interested in its action on members of S_Q. (For degenerate q_i (43) may be adapted by simply omitting those factors in the numerator and denominator for which $q_j = q_i$).

More generally

$$P_Q(\Delta) = \chi_\Delta(Q) \qquad (44)$$

where $\chi_\Delta: \mathbb{R} \to \{0, 1\}$ is defined by

$$\chi_\Delta(x) = 1 \quad \text{if } x \in \Delta$$
$$= 0 \quad \text{if } x \notin \Delta \qquad (45)$$

χ_Δ is the characteristic function which 'picks out' the set Δ.

We can now write our statistical algorithm in the form

$$\text{Prob}(\Delta)_Q^W = \text{Tr}(\chi_\Delta(Q) \cdot W) \qquad (46)$$

So far we have considered Δ as a subset of the spectrum of Q. But clearly (46) also applies where Δ is any subset of the reals. If no element of the spectrum is a member of Δ, then (46) assigns probability zero to this subset. The result (46) can be generalized to

express the joint probability for observing two commuting observables Q and Q' to have values lying in the sets Δ and Δ', in the form
$$\text{Prob}\,(\Delta, \Delta')^W_{Q,Q'} = \text{Tr}\,(\chi_\Delta(Q)\cdot\chi_{\Delta'}(Q')\cdot W) \qquad (47)$$
where the notation on the LHS is a convenient generalization of the symbol $(\Delta)^W_Q$ to express the proposition that measurement results of Q and Q' will lie in the sets Δ and Δ'. We shall discuss later the question of what exactly is meant by the phrase 'simultaneous measurement of two commuting observables'. For the moment we shall use Eq. (47) to *define* the quantum-mechanical joint probability distribution for commuting observables.

We shall now use the result (46) to prove the following theorem.
$$\text{Prob}\,(\Delta)^W_{f(Q)} = \text{Prob}\,(f^{-1}(\Delta))^W_Q \qquad (48)$$
Proof: The LHS $= \text{Tr}(\chi_\Delta(f(Q))\cdot W)$
$\qquad\qquad\quad = \text{Tr}(\chi_{f^{-1}(\Delta)}(Q)\cdot W)$
$\qquad\qquad\quad = \text{RHS}$
where we have used the result
$$\chi_\Delta(f(x)) = \chi_{f^{-1}(\Delta)}(x) \qquad (49)$$
The truth of this property of characteristic functions can be seen by noticing:

If $f(x)\in\Delta$, LHS $= 1$, and then since $x \in f^{-1}(\Delta)$, RHS $= 1$. But if $f(x)\notin\Delta$, LHS $= 0$ and then since $x\notin f^{-1}(\Delta)$, RHS $= 0$. So, for all values of x, the two functions on the left and right hand side of Eq. (49) have the same value. Hence the two functions are equal.

The important result expressed in Eq. (48) is known as the *Statistical Functional Composition Principle*, which we shall abbreviate as STAT FUNC.

STAT FUNC will play an important role in our later discussion of the Kochen–Specker paradox (see Chapter 5). Notice carefully that STAT FUNC does *not* say
$$\text{Prob}\,(f(\Delta))^W_{f(Q)} = \text{Prob}\,(\Delta)^W_Q$$
a result which is in general false when f is not a $1:1$ mapping.

Suppose Q is a maximal observable. By this remember that we mean that the associated self-adjoint operator is nondegenerate—i.e. all its eigenvalues are unequal. Then, if f is a $1:1$ mapping, $f(Q)$ will also be maximal, but if f is a many–one mapping, then $f(Q)$ will be degenerate (nonmaximal).

The Formalism of Quantum Mechanics

Suppose now that Q and R are any two commuting observables. Then there always exists a maximal observable S, such that $Q = f(S)$, $R = g(S)$ for appropriate functions f and g.

We prove this result, confining our attention as usual to the finite-dimensional case.

Proof: Since Q and R commute they can be diagonalized simultaneously. This means that there exists a basis $\{|q_i\rangle\}$ relative to which we can expand

$$Q = \sum_i q_i P_{|q_i\rangle}$$

and such that R can also be expanded in the form

$$R = \sum_i r_i P_{|q_i\rangle}$$

Now write $S = \sum_i s_i P_{|q_i\rangle}$ where all the s_i are different, so that by construction S is maximal.

Then we have

$$Q = f(S)$$
$$R = g(S)$$

where $f: S_S \to S_Q$ and $g: S_S \to S_R$ are defined by

$$f(s_i) = q_i$$
$$g(s_i) = r_i.$$

(Remember that S_Q stands for the spectrum of the observable Q, S_S for the spectrum of the observable S, etc.) So the theorem has been proved by explicit construction.

Note carefully that if $Q = F(R)$ for any observables Q and R and any function F, then Q and R commute, but the converse in general fails—that is to say, if Q and R commute, this does not mean we can write $Q = F(R)$ for some function F. Instead we have the result demonstrated above.

In the special case where either f or g turn out to be 1:1, so that one or other of Q and R is maximal, then we can indeed write the nonmaximal observable as a function of the maximal one. Thus suppose Q is maximal, so f is 1:1. Then f possesses an inverse and we can write

$$S = f^{-1}(Q)$$

So $R = g(S) = g(f^{-1}(Q)) = F(Q)$, where $F = g \circ f^{-1}$.

Notice that if we can write $Q = f(S)$ for any observable Q as a function of some maximal observable S, then the choice of S is by no means unique.

Suppose we can write

$$Q = f(S) = g(R) \qquad (50)$$

for *two* maximal observables S and R. Then if Q is maximal S and R must commute.

Proof: In this case we can write

$$S = f^{-1}(Q) = f^{-1}(g(R))$$

so S is a function of R and hence commutes with R.

But if Q is nonmaximal, then we can always choose an R such that (50) holds where R does not commute with S.

Proof: As before we write $S = \sum_i s_i P_{|q_i\rangle}$, $Q = \sum_i q_i P_{|q_i\rangle}$ where all the s_i are distinct and some of the q_i are equal.

So $Q = f(S)$, where $f(s_i) = q_i$

Suppose without loss of generality that the eigenvalues of Q are enumerated in such a way that $q_1 = q_2 = \alpha$ say. In the space spanned by $|q_1\rangle, |q_2\rangle$ introduce by a unitary transformation a second pair of orthonormal basis vectors which are non-trivial linear combinations of $|q_1\rangle$ and $|q_2\rangle$ and which we label $|q'_1\rangle, |q'_2\rangle$ where these basis vectors are still eigenvectors of Q with the eigenvalues $q'_1 = q'_2 = \alpha$.

Then $Q = q_1 P_{|q_1\rangle} + q_2 P_{|q_2\rangle} + \sum_{i=3}^{N} q_i P_{|q_i\rangle}$

$$= \alpha(P_{|q_1\rangle} + P_{|q_2\rangle}) + \sum_{i=3}^{N} q_i P_{|q_i\rangle}$$

$$= \alpha(P_{|q'_1\rangle} + P_{|q'_2\rangle}) + \sum_{i=3}^{N} q_i P_{|q_i\rangle}$$

$$= g(R)$$

where

$$R = r_1 P_{|q'_1\rangle} + r_2 P_{|q'_2\rangle} + \sum_{i=3}^{N} r_i P_{|q_i\rangle} \qquad (51)$$

and all the $r_i, i = 1 \ldots N$ are distinct so R is maximal and $g: S_R \to S_Q$ is

defined by
$$g(r_1) = \alpha$$
$$g(r_2) = \alpha$$
$$g(r_i) = q_i, \quad i = 3 --- N.$$

Compare the spectral expansion of R with that of S.

$$S = s_1 P_{|q_1\rangle} + s_2 P_{|q_2\rangle} + \sum_{i=3}^{N} s_i P_{|q_i\rangle} \tag{52}$$

From (51) and (52) we see that R and S do not commute, because projection operators such as $P_{|q_1\rangle}$ and $P_{|q'_1\rangle}$, whose ranges are rays which by construction are neither parallel nor orthogonal, do not commute.

(To see this consider $P_{|q_1\rangle} P_{|q'_1\rangle} |q_1\rangle$. The result is a vector parallel to $|q_1\rangle$. But for $P_{|q'_1\rangle} P_{|q_1\rangle} |q_1\rangle$ the result is a vector parallel to $|q'_1\rangle$, so the commutator of the two projection operators $P_{|q'_1\rangle}$ and $P_{|q_1\rangle}$ cannot be identically zero).

So our theorem is proved by direct construction.

As a particular example, suppose $N = 3$ and consider

$$S = s_1 P_1 + s_2 P_2 + s_3 P_3$$
$$R = r_1 P'_1 + r_2 P'_2 + r_3 P_3$$

where P_1, P_2, P_3 project onto the eigenvectors of the maximal observable S and P'_1, P'_2, P_3 onto the eigenvectors of the maximal observable R.

$$\text{Then } P_3 = \chi_{s_3}(S)$$
$$= \chi_{r_3}(R) \tag{53}$$

So the nonmaximal observable P_3 has here been expressed in two different ways as functions of two noncommuting maximal observables.

We illustrate the relationship between the projection operators associated with S and R, here identified with the rays onto which they project (i.e. their ranges), in Fig. 1.

The triad of rays P'_1, P'_2, P_3 is obtained from the triad P_1, P_2, P_3 by 'rotating' the latter triad about the fixed ray P_3. Clearly P_3 can be considered as a member of an infinite number of different triads P'_1, P'_2, P_3, depending on an infinite number of possible rotations about P_3. Each such triad will define a maximal observable (again not

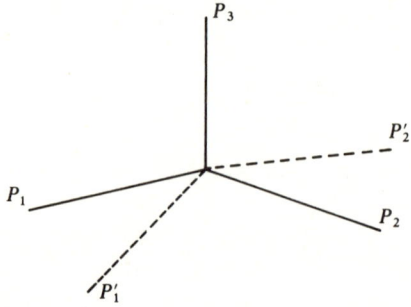

Fig. 1. P_1, P_2, P_3 are a triad of orthogonal rays. The triad P'_1, P'_2, P_3 is obtained from the triad P_1, P_2, P_3 by rotating about the fixed ray P_3.

uniquely) such that P_3 is a function of it, and such that none of these maximal observables commute.

1.4. The Logico-Algebraic Approach

In expounding the von Neumann approach to QM in Section 1.2 we have seen the vital role played by projection operators. They serve both to specify the observables via the spectral expansion (29) and to specify the states, as in (35). The general version of the statistical algorithm (36) involved the more general projection operators $P_Q(\Delta)$ with multi-dimensional ranges. This suggests that we could attempt a more abstract characterization of the formalism of QM by abstracting the essential algebraic structure associated with the collection of all the projection operators on a Hilbert space. Since the range of any projection operator is a subspace, we can set up a 1:1 correspondence between subspaces and projection operators and talk equally about the algebraic structure of the set of all subspaces of Hilbert space. Now this set is partially ordered by the relation of set-theoretic inclusion; more specifically, it is a lattice (see Mathematical Appendix), where the least upper bound of any two subspaces is their *linear span* (the linear span of subspaces A and B will be denoted by $A \oplus B$)—i.e. the subspace comprising all possible linear combinations of vectors from the two subspaces—and the greatest lower bound is the set-theoretic intersection (denoted by $\cap$) of the two subspaces. The lattice is also orthocomplemented if we define the orthocomplement of any subspace as the subspace consisting of all those vectors which are

The Formalism of Quantum Mechanics

orthogonal to all the vectors in the original subspace. We shall use the notation $A^\perp$ to refer to the orthocomplement of A. Reverting to the language of projection operators, we may refer to this algebraic lattice structure as the projection lattice of the Hilbert space.

The first thing to notice about this projection lattice is that it is not a Boolean lattice, because it is not distributive. To see this, consider three unit vectors $|a\rangle$, $|b\rangle$ and $|c\rangle$, where $|c\rangle$ is some non-trivial linear combination of $|a\rangle$ and $|b\rangle$. Denote by $P_{|a\rangle}$ the projection operator whose range is the ray generated by $|a\rangle$. Similarly for $P_{|b\rangle}$ and $P_{|c\rangle}$. Identifying projection operators with the subspaces which comprise their range, we have

$$P_{|c\rangle} \subset P_{|a\rangle} \oplus P_{|b\rangle}.$$

Then

$$P_{|c\rangle} \cap (P_{|a\rangle} \oplus P_{|b\rangle}) \neq (P_{|c\rangle} \cap P_{|a\rangle}) \oplus (P_{|c\rangle} \cap P_{|b\rangle}) \qquad (54)$$

The LHS of (54) is just $P_{|c\rangle}$, while the RHS is $0 \oplus 0 = 0$, where 0 denotes the zero space.

But although the projection lattice is not Boolean, it has plenty of Boolean sublattices. We confine our attention in what follows to Boolean sublattices for which the complement (orthocomplement) coincides with the orthocomplement in the full projection lattice.

Let us take first the case of a two-dimensional Hilbert space $\mathbb{H}_2$. Consider an arbitrary pair of orthogonal unit vectors $|a\rangle$ and $|a'\rangle$ which span the space, and where a and a' may be regarded as the eigenvalues of some observable for which $|a\rangle$ and $|a'\rangle$ are the associated eigenvectors. Then consider the four projection operators 0, $P_{|a\rangle}$, $P_{|a'\rangle}$, and I, where 0 and I project onto the zero space and the whole space respectively. We can represent the sublattice associated with these projection operators by means of a device known to mathematicians as a *Hasse diagram*. Here, elements in a partial ordering are represented by small circles, and an element being 'less than' another element is represented by joining the two elements with a line, the 'smaller' element being placed at the lower end of the line.

The Hasse diagram for the sublattice with elements 0, $P_{|a\rangle}$, $P_{|a'\rangle}$, and I is shown in Fig. 2. It is isomorphic to the familiar Boolean lattice for the power set of the two-element set $\{a, a'\}$. The Hasse diagram for the power set lattice is shown in Fig. 3.

But of course there are lots of other one-dimensional subspaces apart from those specified by $P_{|a\rangle}$ and $P_{|a'\rangle}$. Other orthogonal pairs,

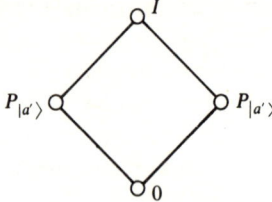

Fig. 2. Hasse diagram for the sublattice with elements 0, $P_{|a\rangle}$, $P_{|a'\rangle}$, and I.

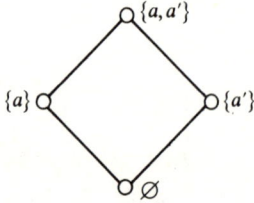

Fig. 3. Hasse diagram for the Boolean power set lattice of the two-element set $\{a, a'\}$.

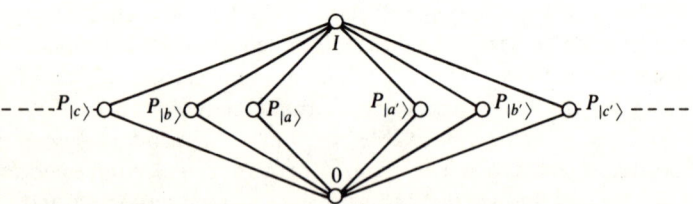

Fig. 4. Projection lattice for a two-dimensional Hilbert space. The elements associated with orthogonal pairs of projections $P_{|a\rangle}$ and $P_{|a'\rangle}$, $P_{|b\rangle}$ and $P_{|b'\rangle}$, $P_{|c\rangle}$ and $P_{|c'\rangle}$ are illustrated.

$P_{|b\rangle}$, $P_{|b'\rangle}$; $P_{|c\rangle}$, $P_{|c'\rangle}$, and so on, could be considered. The projection lattice for $\mathbb{H}_2$ would look like Fig. 4, where we have just shown the elements $P_{|a\rangle}$, $P_{|a'\rangle}$, $P_{|b\rangle}$, $P_{|b'\rangle}$, $P_{|c\rangle}$, $P_{|c'\rangle}$.

The whole lattice is non-Boolean, as we have seen, but it can be regarded as an infinite collection of interlocking Boolean sublattices. However, these Boolean sublattices only interlock in a trivial way— viz. they all possess 0 and I in common. When we turn to three-

The Formalism of Quantum Mechanics

dimensional and higher-dimensional Hilbert spaces, the way in which the Boolean sublattices interlock becomes highly nontrivial.

Let us consider a typical Boolean sublattice for a three-dimensional Hilbert space. Take three orthogonal unit vectors $|a\rangle$, $|b\rangle$, and $|c\rangle$ with associated projection operators $P_{|a\rangle}$, $P_{|b\rangle}$, and $P_{|c\rangle}$, and consider the following set of eight projection operators: 0, $P_{|a\rangle}$, $P_{|b\rangle}$, $P_{|c\rangle}$, $P_{|a\rangle} + P_{|b\rangle}$, $P_{|a\rangle} + P_{|c\rangle}$, $P_{|b\rangle} + P_{|c\rangle}$, and $P_{|a\rangle} + P_{|b\rangle} + P_{|c\rangle} = I$. The Hasse diagram looks like Fig. 5. Again we may regard a, b, and c as the eigenvalues of some observable with the eigenvectors $|a\rangle$, $|b\rangle$, and $|c\rangle$, and the above lattice is isomorphic to the power set Boolean lattice for the set of eigenvalues $\{a, b, c\}$ which is illustrated in Fig. 6.

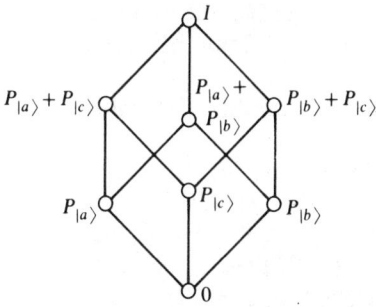

Fig. 5. Boolean sublattice for a three-dimensional Hilbert space, where $|a\rangle$, $|b\rangle$, and $|c\rangle$ are three orthogonal unit vectors.

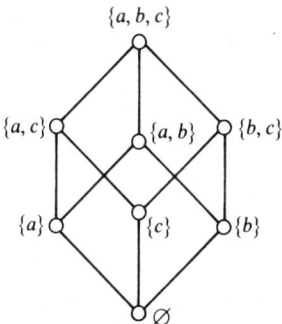

Fig. 6. Hasse diagram for the Boolean power set lattice of the three-element set $\{a, b, c\}$.

Notice that the Boolean sublattice shown in Fig. 5 is maximal in the sense that there is no other Boolean sublattice of which it is a sublattice.

As we 'rotate' the triad of orthonormal vectors $|a\rangle$, $|b\rangle$, and $|c\rangle$ into all possible orientations, an infinite number of such Boolean sublattices are generated; but they no longer overlap just in sharing 0 and I. For example, all the lattices generated by triads of orthogonal vectors obtained by 'rotating' the triad $|a\rangle$, $|b\rangle$, and $|c\rangle$ about the direction $|c\rangle$ will have the projection operator $P_{|c\rangle}$ in common. So the Boolean sublattices now interlock in a highly nontrivial fashion. Clearly this will also be the case for any dimension greater than three. We shall find later that two-dimensional Hilbert spaces, which describe the quantum mechanics of the spin properties of spin-$\frac{1}{2}$ particles, are quite atypical and allow hidden variable reconstructions for the QM statistics. This is ultimately related to the peculiarly simple character of the associated projection lattice.

We have seen, then, that the essential mathematical structure for the QM formalism is the projection lattice of the relevant Hilbert space. This suggests the programme of characterizing the projection lattice axiomatically independently of its origin in Hilbert space. Such axiomatizations are known as the logico-algebraic approach to QM. The reason why it is called 'algebraic' is clear. The connection with logic will be developed in more detail in Chapter 7. Briefly, one is appealing to well-known connections between Boolean lattices and classical propositional logic. The idea is that the more general non-Boolean lattices contemplated in QM are similarly related to a new generalized logic, commonly called quantum logic.

But one object of the logico-algebraic axiomatizations is to allow variants which are not equivalent to the standard Hilbert space approach, for example, to characterize lattices—or even more general posets—which are not the projection lattice of a Hilbert space over the complex field. This programme is discussed in Jauch (1968). In the present work we shall not be concerned with attempts to vary or generalize QM, and we shall treat the logico-algebraic approach as just another way of looking at the standard von Neumann Hilbert space formalism.

A word of warning: do not confuse the logico-algebraic approach to QM with the quite different algebraic approach. This abstracts from the algebra of linear operators on a Hilbert space which form what mathematicians call a *Banach algebra*. A Banach algebra has both

The Formalism of Quantum Mechanics

topological and algebraic aspects, and in this sense is a richer structure than the projection lattice, which is purely algebraic. Again the algebraic approach can be developed axiomatically and used to vary or generalize the standard Hilbert space theory from which it originates. In the present work we shall not have occasion to refer to this algebraic approach to QM, nor to some of the other standard approaches described in the physics literature, such as S-matrix theory, the Feynman path-integral approach, and second quantization.

1.5. Gleason's Theorem

We want to explain what is meant by a *probability measure* defined over the projection lattice of a Hilbert space. By this we will mean a map μ from the projection operators into the real numbers in the closed interval $[0, 1]$, which satisfy the usual requirements for a probability measure on each maximal Boolean sublattice. This just amounts to requiring that

$$\mu\left(\sum_i P_i\right) = \sum_i \mu(P_i) \tag{55}$$

and

$$\mu(0) = 0, \quad \mu(I) = 1 \tag{56}$$

where $\{P_i\}$ is any countable set of mutually orthogonal projection operators. In the case of finite-dimensional Hilbert spaces, the number of mutually orthogonal projection operators is of course finite (indeed, never greater than the dimension of the Hilbert space); so in this case the countable additivity expressed by the general requirement (55) reduces always just to finite additivity.

In particular, if the map μ assigns the numbers $\mu(P_{|q_i\rangle})$ to the projection operators occurring in the spectral expansion of an operator Q, as in Eq. (29), then we require

$$\sum_{i=1}^{N} \mu(P_{|q_i\rangle}) = \mu\left(\sum_{i=1}^{N} P_{|q_i\rangle}\right) = \mu(I) = 1 \tag{57}$$

The question now arises: what are the possible probability measures definable over the projection lattice of a Hilbert space? We already have at hand a class of such measures. Thus

$$\mu(P_i) = \mathrm{Tr}\,(P_i \cdot W) \tag{58}$$

where W is a statistical operator as specified in Eq. (35), certainly satisfies the conditions we have imposed. Remember that

$$\text{Tr}(W) = \sum_k w_k \text{Tr}(P_{|\psi_k\rangle}) = \sum_k w_k = 1 \qquad (59)$$

since $\text{Tr}(P_{|\psi_k\rangle}) = 1$.

In 1957 Gleason proved a very remarkable converse of this result. All probability measures over the projection lattice are of the form (58), provided that the dimension of the Hilbert space is greater than two. Again, the somewhat anomalous status of the two-dimensional case is brought out in Gleason's theorem. We shall not give a proof of Gleason's theorem, although it provides the key to the Kochen–Specker paradox discussed in Chapter 5. The reason for this is that we shall prove directly the relevant corollary of Gleason's theorem which is of importance here.

We make, however, a few comments on the significance of the measures specified by (58). Let us consider, to be more specific, a real three-dimensional Hilbert space. This is just the familiar Euclidean 3-space. The projection operators with one-dimensional range correspond just to lines through the origin in all possible orientations. Each line will intersect the unit sphere in two antipodal points. We can now think of the measure μ as inducing a map μ' into $[0, 1]$ of the points on the surface of the unit sphere, where antipodal points are always given the same value. The remarkable thing about μ' is that it is continuous as we move around the unit sphere. Informally, sufficiently small changes in position on the sphere produce only small changes in the associated probability number. Some probability measures are ruled out on the basis of Gleason's theorem—simply because they are discontinuous.

Suppose we tried the following map for the unit sphere in Euclidean 3-space. For each orthogonal triad of directions, give *one* direction the value one and both others zero. The values for each triad then certainly sum to one; but because of the complicated way in which all the possible triads of directions interlock (as discussed in the previous section) this construction is simply not possible. One way of seeing this is just to notice that such a map would have to be discontinuous, and Gleason's theorem assures us that discontinuous measures are not possible for the three-dimensional case.

Why must the map we have introduced be discontinuous? Informally we proceed as follows.

The Formalism of Quantum Mechanics

Consider two points p and q on the unit sphere such that $\mu'(p) = 1$ and $\mu'(q) = 0$. Join pq by the shorter arc of a great circle. Choose any point r between p and q. Then $\mu'(r)$ is either 0 or 1. If it is 1 consider the arc rq, if it is 0 consider the arc pr. In either case we have reduced the (geodesic) distance between points with values 0 and 1. The process can be continued indefinitely, so that there is in fact *no* least distance between points of probability assignment 0 and 1. So the probability assignments must jump by a finite amount, viz. 1, in an infinitesimal distance, so the map must be discontinuous.

Since the result is of some importance, we shall now give a more rigorous proof, which shows that the result is topological rather than metrical. The surface of the unit sphere is a connected manifold. By a 'manifold' one means here that it is locally homeomorphic to the Euclidean plane. A connected manifold is one which cannot be represented as the union of disjoint open sets. The open sets on the manifold are just the images under the homeomorphism mentioned of the open sets defined in the Euclidean plane equipped with the usual 'ε-neighbourhood' topology.

Consider now the map μ' specialized to the case we are considering, where the range is the two-element set $\{0, 1\}$. Suppose the map is continuous. Then consider the inverse images of the open sets $\{0\}$ and $\{1\}$ in the range of the map. These will themselves be open sets on the surface of the sphere. (Remember the topological definition of continuity—the inverse image of every open set in the range of a continuous map is an open set of the domain.) But they will also satisfy

$$\mu'^{-1}(\{1\}) = \mathscr{C}\mu'^{-1}(\{0\})$$

where $\mathscr{C}$ denotes the set-theoretic complement. This follows since *every* point on the sphere is mapped onto either 1 or 0.

So if we denote the whole sphere by S

$$S = \mu'^{-1}(\{0\}) \cup \mathscr{C}\mu'^{-1}(\{0\})$$
(by the definition of complement)
$$= \mu'^{-1}(\{0\}) \cup \mu'^{-1}(\{1\})$$

and this represents the surface of the sphere as the union of two disjoint open sets, which contradicts the topological property of the surface of the sphere of being connected. Hence, by *reductio ad absurdum*, we have demonstrated that the map cannot be continuous.

We conclude this section with a note for those having some familiarity with measure theory. Gleason's theorem is really a generalized *Radon–Nikodym* theorem, which generates the Gleason measures from the canonical measure Trace. Trace is the measure which assigns to each projection operator just the dimension of its range. This can be normalized by dividing by the dimension N of the Hilbert space (in the case of finite-dimensional Hilbert spaces). This canonical measure is obtained from (58) by taking W as a 'constant', $1/N \times I$, where I is the identity operator. In the more general case given by Gleason's result with W *not* a 'constant', then W plays the role of a sort of Radon–Nikodym derivative of the Gleason measure with respect to the canonical measure, integration being replaced by diagonal sum (the trace operation).

1.6. The Quantum Mechanics of Many-Particle Systems

The mathematical machinery for describing the states of many-particle systems in QM is provided by the notion of the tensor product (for details see Mathematical Appendix, pp. 174 ff.). Considering the case of two particles, whose individual states are represented as unit vectors in Hilbert spaces $\mathbb{H}_1$ and $\mathbb{H}_2$, then the Hilbert space appropriate to representing the states of the composite system is the tensor product of $\mathbb{H}_1$ and $\mathbb{H}_2$, written $\mathbb{H}_1 \otimes \mathbb{H}_2$. It is important to notice that vectors in $\mathbb{H}_1 \otimes \mathbb{H}_2$ are in general linear combinations of the tensor products of vectors in $\mathbb{H}_1$ and $\mathbb{H}_2$; only in special cases can they be written as a simple tensor product. Be careful not to confuse the tensor product of vector spaces with the tensor product of vectors belonging to the spaces.

Suppose we have an observable Q_1 associated with the particle 1, whose states are described by vectors in the Hilbert space $\mathbb{H}_1$, and associated with the self-adjoint operator which we also write as Q_1. Then we can also think of Q_1 as an observable relating to the composite system of particles 1 and 2. The self-adjoint operator associated with Q_1, when viewed as an observable for the composite system, is $Q_1 \otimes I$, where I denotes the identity operator on $\mathbb{H}_2$. Similarly, an observable R_2 pertaining to particle 2 is represented on the product space by $I \otimes R_2$, where I now denotes the identity operator on $\mathbb{H}_1$. We are here using still another sense of tensor product, the tensor product of linear operators. We shall use the convention that in the tensor product $A \otimes B$ of two operators, the

The Formalism of Quantum Mechanics 31

left-hand operator in the product shall refer to $\mathbb{H}_1$ and the right-hand operator to $\mathbb{H}_2$.

Since $I \otimes R_2$ commutes with $Q_1 \otimes I$, we can ask for the joint probability distribution that Q_1 has the value q_i and R_2 the value r_j, where q_i is the ith eigenvalue of Q_1 and r_j the jth eigenvalue of R_2.

Utilizing (47), it is given by

$$\begin{aligned}\text{Prob}\,(q_i, r_j)^W_{Q_1, R_2} &= \text{Tr}\,(\chi_{q_i}(Q_1 \otimes I) \cdot \chi_{r_j}(I \otimes R_2) \cdot W) \\ &= \text{Tr}\,((\chi_{q_i}(Q_1) \otimes I) \cdot (I \otimes \chi_{r_j}(Q_2)) \cdot W) \\ &= \text{Tr}\,((\chi_{q_i}(Q_1) \otimes \chi_{r_j}(R_2)) \cdot W) \\ &= \text{Tr}\,((P_{Q_1}(q_i) \otimes P_{R_2}(r_j)) \cdot W) \end{aligned} \quad (60)$$

where W is the statistical operator for the joint system.

We consider briefly how to represent this tensor product idea using the Dirac notation. The tensor product of $|\alpha\rangle$ and $|\beta\rangle$ is written as $|\alpha\rangle|\beta\rangle \underset{Df}{=} |\alpha\rangle \otimes |\beta\rangle$. Similarly the tensor product of the bras $\langle\alpha|$ and $\langle\beta|$ is defined thus: $(\langle\alpha| \otimes \langle\beta|)(|\gamma\rangle \otimes |\delta\rangle) \underset{Df}{=} \langle\alpha|\gamma\rangle \cdot \langle\beta|\delta\rangle$ for arbitrary kets $|\gamma\rangle$ and $|\delta\rangle$.

In abbreviated notation we write $\langle\alpha| \otimes \langle\beta|$ as $\langle\alpha|\langle\beta|$.

We can now understand how to write

$$P_{|q_i\rangle} \otimes P_{|r_j\rangle} = |q_i\rangle|r_j\rangle\langle q_i|\langle r_j| \quad (61)$$

To avoid confusion in forming inner products of tensor products, it is best to use the functional notation $\langle\alpha|\langle\beta|\,(|\gamma\rangle|\delta\rangle)$, which is defined as above to mean $\langle\alpha|\gamma\rangle \cdot \langle\beta|\delta\rangle$. In particular, if $|\Psi\rangle$ is a vector in the tensor product space and $W = P_{|\Psi\rangle}$, then

$$\text{Prob}\,(q_i, r_j)^W_{Q_1, R_2} = \sum_{s|q_s = q_i} \sum_{t|r_t = r_j} |\langle q_s|\langle r_t|(|\Psi\rangle)|^2 \quad (62)$$

where the summations allow as usual for degeneracy. Dirac himself writes $\langle q_s|\langle r_t| \underset{Df}{=} \langle q_s r_t|$ where the order of the eigenvalues q_s, r_t on the RHS of this relation is crucial; the first number refers to particle 1 and the second to particle 2. With this abbreviated notation we can get rid of the round brackets in Eq. (62) and write simply

$$\text{Prob}\,(q_i, r_j)^W_{Q_1, R_2} = \sum_{s|q_s = q_i} \sum_{t|r_t = r_j} |\langle q_s r_t|\Psi\rangle|^2$$

Dirac writes $|\alpha\rangle|\beta\rangle \underset{Df}{=} |\alpha\beta\rangle$ where $|\alpha\rangle$ is any ket in H_1 and $|\beta\rangle$ any ket in H_2, corresponding to $\langle\alpha|\langle\beta| = \langle\alpha\beta|$ for bras. So $\langle\alpha|\langle\beta|(|\gamma\rangle|\delta\rangle)$ is written as $\langle\alpha\beta|\gamma\delta\rangle \underset{Df}{=} \langle\alpha|\gamma\rangle \cdot \langle\beta|\delta\rangle$. This abbreviated notation is often convenient, but on the whole we recommend the more transparent functional notation when dealing with inner products of tensor products.

1.7. Angular Momentum in Quantum Mechanics

In the following chapters we shall often use as examples the finite-dimensional Hilbert spaces required for describing the spin angular momentum of quantum-mechanical systems. We shall collect together in this section all the results we shall use later in discussing the properties of spin-$\frac{1}{2}$ and spin-1 systems. Spin in QM is a vector quantity **S** associated with the 'internal' degrees of freedom of a system. We denote the observables corresponding to the X-, Y-, and Z-components of spin relative to a Cartesian reference frame by S_x, S_y, and S_z respectively. The magnitude of the spin vector is denoted by S.

So we have the basic relationship

$$S^2 = S_x^2 + S_y^2 + S_z^2 \tag{63}$$

We assume that the observables S_x, S_y, S_z (more correctly their associated self-adjoint operators) obey the same commutation relations as those applying in the case of 'orbital' angular momentum (i.e. angular momentum defined as $\mathbf{r} \times \mathbf{p}$ in the usual way). So we have the commutation relations already given in Section 1.1 (see Eq. (13)):

$$\left. \begin{array}{l} S_x S_y - S_y S_x = iS_z \\ S_y S_z - S_z S_y = iS_x \\ S_z S_x - S_x S_z = iS_y \end{array} \right\} \tag{64}$$

where we choose units so that Planck's reduced constant $\hbar$ is equal to unity.

The problem of determining the eigenvalues of S^2, S_x, S_y and S_z is a standard one, treated in nearly all textbooks on QM. We shall simply state the results.

The eigenvalues of S^2 are $s(s+1)$ where s is an integer or half-integer. For a given value of s, S^2 has a $(2s+1)$-fold degeneracy which can be removed by specifying the eigenvalue of any one of the components of the vector **S**. The eigenvalues for these components are

The Formalism of Quantum Mechanics

simply m where m has the value $-s, -s+1 \ldots s-1, s$, i.e. m has any one of $2s+1$ values spaced at integral intervals between $-s$ and $+s$.

We denote the eigenvector of S_z belonging to the eigenvalue m, for a fixed s, by $|S_z = m\rangle$; similarly, we use the notation $|S_x = m\rangle$ and $|S_y = m\rangle$ to denote eigenvectors of S_x and S_y associated with the eigenvalue m.

Discussion of angular momentum in QM is greatly facilitated by introducing the 'raising' and 'lowering' operators S_+, S_- which are defined by

$$S_+ = S_x + iS_y \\ S_- = S_x - iS_y \quad (65)$$

So we can write

$$S_x = \tfrac{1}{2}(S_+ + S_-) \\ S_y = \frac{1}{2i}(S_+ - S_-) \quad (66)$$

The names 'raising' and 'lowering' operators derive from the following results showing the effect of operating with $S_\pm$ on $|S_z = m\rangle$. S_+ produces the ket $|S_z = m+1\rangle$ to within an appropriate factor. Hence it is called a raising operator, because it raises the eigenvalue of S_z from m to $m+1$. Similarly, S_- produces the ket $|S_z = m-1\rangle$ to within a multiplying factor. So S_- is a lowering operator because it lowers the eigenvalue of S_z from m to $m-1$.

More specifically we have

$$S_\pm |S_z = m\rangle = C_\pm(s,m)|S_z = m \pm 1\rangle \quad (67)$$

where

$$C_\pm(s,m) = s(s+1) - m(m \pm 1) \quad (68)$$

Notice that when $m = s$

$$C_+(s,s) = 0 \quad (69)$$

and when $m = -s$

$$C_-(s,-s) = 0 \quad (70)$$

This means that acting repeatedly with S_+ on $|S_z = m\rangle$ cannot increase the eigenvalue of S_z above the value $+s$. Similarly S_- acting repeatedly on $|S_z = m\rangle$ cannot reduce the eigenvalue of S_z below the value $-s$.

We shall be interested in particular in the two special cases $s = \tfrac{1}{2}$ and $s = 1$.

$s = \frac{1}{2}$

The possible values of m are now $\pm\frac{1}{2}$.

It is convenient in this case to define a new operator

$$\sigma = 2 \times \mathbf{S} \tag{71}$$

so the eigenvalues of $\sigma_x, \sigma_y, \sigma_z$ are ± 1.

We introduce the notation

$$\left.\begin{array}{l}|\alpha\rangle = |\sigma_z = +1\rangle \\ |\beta\rangle = |\sigma_z = -1\rangle\end{array}\right\} \tag{72}$$

So

$$\left.\begin{array}{l}\sigma_z|\alpha\rangle = |\alpha\rangle \\ \sigma_z|\beta\rangle = -|\beta\rangle\end{array}\right\} \tag{73}$$

Since $C_+(\frac{1}{2}, -\frac{1}{2}) = C_-(\frac{1}{2}, \frac{1}{2}) = 1$ it follows immediately from (66), (67), and (71) that

$$\left.\begin{array}{l}\sigma_x|\alpha\rangle = |\beta\rangle \\ \sigma_x|\beta\rangle = |\alpha\rangle\end{array}\right\} \tag{74}$$

and

$$\left.\begin{array}{l}\sigma_y|\alpha\rangle = i|\beta\rangle \\ \sigma_y|\beta\rangle = -i|\alpha\rangle\end{array}\right\} \tag{75}$$

$\sigma_x, \sigma_y, \sigma_z$ are known as the Pauli spin operators. In the representation afforded by the basis vectors $|\alpha\rangle$ and $|\beta\rangle$ we obtain the familiar matrices

$$\left.\begin{array}{l}\sigma_x = \begin{pmatrix} 0 & 1 \\ 1 & 0 \end{pmatrix} \\ \sigma_y = \begin{pmatrix} 0 & -i \\ i & 0 \end{pmatrix} \\ \sigma_z = \begin{pmatrix} 1 & 0 \\ 0 & -1 \end{pmatrix}\end{array}\right\} \tag{76}$$

We notice that

$$\sigma_x^2 = \sigma_y^2 = \sigma_z^2 = I \tag{77}$$

We consider now the problem of finding the eigenvalues and eigenvectors of the spin component $\boldsymbol{\sigma} \cdot \mathbf{n}$ along some direction specified by the unit vector $\mathbf{n}$, which may be different from the Z-axis.

To simplify the discussion, choose axes X, Y, and Z so that $\mathbf{n}$ lies in the XZ plane and makes an angle θ with the Z-axis. Then $\boldsymbol{\sigma} \cdot \mathbf{n}$

The Formalism of Quantum Mechanics

$= \sigma_z \cos\theta + \sigma_x \sin\theta$. From (76) the matrix representation of $\boldsymbol{\sigma}\cdot\mathbf{n}$ in the basis provided by $|\alpha\rangle$ and $|\beta\rangle$ is then

$$\boldsymbol{\sigma}\cdot\mathbf{n} = \begin{pmatrix} \cos\theta & \sin\theta \\ \sin\theta & -\cos\theta \end{pmatrix} \tag{78}$$

We consider the eigenvalue problem

$$\begin{pmatrix} \cos\theta & \sin\theta \\ \sin\theta & -\cos\theta \end{pmatrix}\begin{pmatrix} u \\ v \end{pmatrix} = \lambda \begin{pmatrix} u \\ v \end{pmatrix} \tag{79}$$

where we wish to find the permitted values of the eigenvalue λ, and the associated expansion coefficients u and v for the eigenvectors in terms of $|\alpha\rangle$ and $|\beta\rangle$.

The equations for determining u and v are just

$$\left.\begin{array}{l}\cos\theta\cdot u + \sin\theta\cdot v = \lambda u \\ \sin\theta\cdot u - \cos\theta\cdot v = \lambda v\end{array}\right\} \tag{80}$$

Consistency of these equations requires

$$(\lambda - \cos\theta)(\lambda + \cos\theta) = \sin^2\theta$$
$$\text{or} \quad \lambda^2 = \cos^2\theta + \sin^2\theta = 1$$

So

$$\lambda = \pm 1 \tag{81}$$

For $\lambda = +1$ we find from (80)

$$u/v = \cos(\theta/2)/\sin(\theta/2).$$

For $\lambda = -1$

$$u/v = -\sin(\theta/2)/\cos(\theta/2).$$

Hence we obtain for the normalized eigenvalues of $\boldsymbol{\sigma}\cdot\mathbf{n}$ the results

$$\left.\begin{array}{l}|\boldsymbol{\sigma}\cdot\mathbf{n} = +1\rangle = \sin\theta/2\,|\beta\rangle + \cos\theta/2\,|\alpha\rangle \\ |\boldsymbol{\sigma}\cdot\mathbf{n} = -1\rangle = \cos\theta/2\,|\beta\rangle - \sin\theta/2\,|\alpha\rangle\end{array}\right\} \tag{82}$$

As a particular case, take $\mathbf{n}$ along the positive X-axis, and denoting $|\sigma_x = +1\rangle$ by $|\gamma\rangle$ and $|\sigma_x = -1\rangle$ by $|\delta\rangle$, we obtain the simple relations

$$\left.\begin{array}{l}|\gamma\rangle = \dfrac{1}{\sqrt{2}}(|\beta\rangle + |\alpha\rangle) \\ \\ |\delta\rangle = \dfrac{1}{\sqrt{2}}(|\beta\rangle - |\alpha\rangle)\end{array}\right\} \tag{83}$$

which we can equally well solve for $|\alpha\rangle$ and $|\beta\rangle$ to give

$$\left.\begin{array}{l}|\alpha\rangle = \dfrac{1}{\sqrt{2}}(|\gamma\rangle - |\delta\rangle) \\[1em] |\beta\rangle = \dfrac{1}{\sqrt{2}}(|\gamma\rangle + |\delta\rangle)\end{array}\right\} \quad (84)$$

In passing, we note the connection between the spin operators and rotations. A rotation of the coordinate system through an angle θ in a clockwise sense about the unit vector $\mathbf{n}$ is effected by the unitary operator $e^{-i(\mathbf{S}\cdot\mathbf{n})\theta}$. Now $\sigma_z \cos\theta + \sigma_x \sin\theta$ is obtained from σ_z by a rotation of the Z-axis through an angle θ about the positive Y-axis. The corresponding rotation operator is

$$R(\theta) = e^{-i\cdot\sigma_y\cdot\frac{1}{2}\theta} \quad (85)$$

Expanding in powers of θ and remembering $\sigma_y^2 = I$ gives

$$R(\theta) = \cos(\theta/2)\cdot I - i\sin(\theta/2)\cdot\sigma_y \quad (86)$$

We can now easily verify that the eigenvectors given in (82) are simply $R(\theta)|\alpha\rangle$ and $R(\theta)|\beta\rangle$.

$s = 1$

The possible values of m in this case are $-1, 0$ and $+1$.

Using Eqs. (66), (67), and (68), we can easily construct the results of operating with S_x, S_y, and S_z on the kets $|S_z = -1\rangle$, $|S_z = 0\rangle$, and $|S_z = +1\rangle$. We find in this way the following matrix representation of the operators S_x, S_y, and S_z in the basis afforded by the kets $|S_z = -1\rangle$, $|S_z = 0\rangle$, and $|S_z = +1\rangle$.

$$\left.\begin{array}{l} S_x = \dfrac{1}{\sqrt{2}}\begin{pmatrix} 0 & 1 & 0 \\ 1 & 0 & 1 \\ 0 & 1 & 0 \end{pmatrix} \\[2em] S_y = \dfrac{1}{\sqrt{2}}\begin{pmatrix} 0 & i & 0 \\ -i & 0 & i \\ 0 & -i & 0 \end{pmatrix} \\[2em] S_z = \dfrac{1}{\sqrt{2}}\begin{pmatrix} -1 & 0 & 0 \\ 0 & 0 & 0 \\ 0 & 0 & 1 \end{pmatrix}\end{array}\right\} \quad (87)$$

The Formalism of Quantum Mechanics

It is easily verified by 'brute force' that the matrices S_x^2, S_y^2, and S_z^2 all commute for this particular case $s = 1$.

A more elegant way of proving this important result is to note the identity

$$[S_x^2, S_y^2] = (S_z - 1)S_+^2 - (S_z + 1)S_-^2 \tag{88}$$

The RHS of (88) is clearly zero when acting on any state vector in the space spanned by the kets $|S_z = -1\rangle$, $|S_z = 0\rangle$, and $|S_z = +1\rangle$.

Since the X- and Y-axes can be chosen along any two arbitrary orthogonal directions, we can conclude that similarly

$$[S_x^2, S_z^2] = [S_y^2, S_z^2] = 0.$$

We now show that the kets $|S_x = 0\rangle$, $|S_y = 0\rangle$, and $|S_z = 0\rangle$ provide an orthonormal basis which simultaneously diagonalizes S_x^2, S_y^2, and S_z^2. Using Eqs. (66), (67), and (68) we can easily check the following results

$$|S_x = 0\rangle = \frac{1}{\sqrt{2}}(|S_z = -1\rangle - |S_z = +1\rangle)$$

$$|S_y = 0\rangle = \frac{1}{\sqrt{2}}(|S_z = -1\rangle + |S_z = +1\rangle) \tag{89}$$

Clearly these two kets are orthogonal to each other and to $|S_z = 0\rangle$.

The following results are now easily obtained

$$S_x^2 \cdot \begin{cases} |S_x = 0\rangle = 0 \\ |S_y = 0\rangle = |S_y = 0\rangle \\ |S_z = 0\rangle = |S_z = 0\rangle \end{cases}$$

$$S_y^2 \cdot \begin{cases} |S_x = 0\rangle = |S_x = 0\rangle \\ |S_y = 0\rangle = 0 \\ |S_z = 0\rangle = |S_z = 0\rangle \end{cases} \tag{90}$$

$$S_z^2 \cdot \begin{cases} |S_x = 0\rangle = |S_x = 0\rangle \\ |S_y = 0\rangle = |S_y = 0\rangle \\ |S_z = 0\rangle = 0 \end{cases}$$

Hence in the basis $\{|S_x = 0\rangle, |S_y = 0\rangle, |S_z = 0\rangle\}$ we have the following diagonal representations of $S_x^2, S_y^2,$ and S_z^2

$$S_x^2 = \begin{pmatrix} 0 & & \\ & 1 & \\ & & 1 \end{pmatrix}$$

$$S_y^2 = \begin{pmatrix} 1 & & \\ & 0 & \\ & & 1 \end{pmatrix} \qquad (91)$$

$$S_z^2 = \begin{pmatrix} 1 & & \\ & 1 & \\ & & 0 \end{pmatrix}$$

As a check, we notice at once that

$$S_x^2 + S_y^2 + S_z^2 = \begin{pmatrix} 2 & & \\ & 2 & \\ & & 2 \end{pmatrix}$$

as we expect for a spin-1 system.

Consider now the observable

$$H_s = a\, S_x^2 + b\, S_y^2 + c\, S_z^2 \qquad (92)$$

where a, b, and c are three distinct real numbers. In the basis $\{|S_x = 0\rangle, |S_y = 0\rangle, |S_z = 0\rangle\}$, H_s is represented by the diagonal matrix

$$\begin{pmatrix} b+c & & \\ & a+c & \\ & & a+b \end{pmatrix} \qquad (93)$$

So the eigenvalues of H_s are the numbers $b+c$, $a+c$, and $a+b$, which are themselves distinct if a, b, and c are distinct; so H_s is a maximal operator, unlike $S_x^2, S_y^2,$ and S_z^2 individually, which are clearly degenerate.

Denoting the eigenkets of H_s by the abbreviated notation $|b+c\rangle, |a+c\rangle,$ and $|a+b\rangle$, we have then the simple results

$$\begin{aligned} |b+c\rangle &= |S_x = 0\rangle \\ |a+c\rangle &= |S_y = 0\rangle \\ |a+b\rangle &= |S_z = 0\rangle \end{aligned} \qquad (94)$$

Using (89) and (94), we can express the basis $\{|S_z = -1\rangle, |S_z = 0\rangle,$

$|S_z = +1\rangle\}$ in terms of the basis $\{|b+c\rangle, |a+c\rangle, |a+b\rangle\}$ in the following way

$$\left.\begin{array}{l} |S_z = -1\rangle = \dfrac{1}{\sqrt{2}}(|a+c\rangle + |b+c\rangle) \\[6pt] |S_z = 0\rangle = |a+b\rangle \\[6pt] |S_z = +1\rangle = \dfrac{1}{\sqrt{2}}(|a+c\rangle - |b+c\rangle) \end{array}\right\} \quad (95)$$

The properties of the spin Hamiltonian H_s, introduced in a purely formal way above, will be used extensively in discussing the Kochen–Specker paradox in Chapters 5 and 6 below. For the moment we shall merely note that S_x^2, S_y^2, and S_z^2 can all be expressed as functions of H_s in the following way:

$$S_x^2 = (a-b)^{-1}(c-a)^{-1}[H_s - (b+c)][H_s - 2a] \quad (96)$$

$$S_y^2 = (b-c)^{-1}(a-b)^{-1}[H_s - (c+a)][H_s - 2b] \quad (97)$$

$$S_z^2 = (c-a)^{-1}(b-c)^{-1}[H_s - (a+b)][H_s - 2c] \quad (98)$$

S_x^2, S_y^2, and S_z^2 are in fact projection operators whose ranges are the two-dimensional subspaces spanned by the pairs $\{|a+c\rangle, |a+b\rangle\}$, $\{|b+c\rangle, |a+b\rangle\}$, and $\{|b+c\rangle, |a+c\rangle\}$ respectively.

We can introduce projection operators defined by

$$\left.\begin{array}{l} P_x = I - S_x^2 \\ P_y = I - S_y^2 \\ P_z = I - S_z^2 \end{array}\right\} \quad (99)$$

whose ranges are the rays associated with $|b+c\rangle$, $|a+c\rangle$, and $|a+b\rangle$ respectively.

Note also that

$$H_s = (b+c)P_x + (a+c)P_y + (a+b)P_z \quad (100)$$

which is just the spectral expansion for H_s, and now of course

$$P_x + P_y + P_z = 3 \cdot I - (S_x^2 + S_y^2 + S_z^2) = 3 \cdot I - 2 \cdot I = I$$

in the usual way.

1.8. Angular Momentum of a Composite System

For a two-particle system, we have seen in Section 1.6 that we have to employ a tensor product space to describe the states of the composite system. We shall now consider the problem of finding the eigenvalues and eigenkets for the total spin angular momentum **S** of two particles in terms of the eigenvalues and eigenkets of the spin angular momenta **S**$_1$ and **S**$_2$ of the component particles separately. The general result for solving this problem is well known. If s_1 and s_2 are the spin quantum numbers for the two particles, then the total spin quantum number s for the combined system can range in integral steps from $|s_1 - s_2|$ to $s_1 + s_2$. For a given s, the eigenvalues of the Z-component of total spin S_z can range in integral steps from $-s$ to $+s$ in the usual way. The problem of expressing $|S_z = m\rangle$ for a given permitted s in terms of $|S_{1z} = m'\rangle$ and $|S_{2z} = m''\rangle$, where m' runs from $-s_1$ to $+s_1$ and m'' from $-s_2$ to $+s_2$, is solved in terms of the so-called Clebsch–Gordon coefficients defined by

$$|S_z = m\rangle = \sum_{m'=-s_1}^{s_1} \sum_{m''=-s_2}^{s_2} c(m', m''; m, s, s_1, s_2)$$
$$|S_{1z} = m'\rangle |S_{2z} = m''\rangle \qquad (101)$$

The coefficients $c(m', m''; m, s, s_1, s_2)$ are in general rather complicated functions of their arguments, but for low values of s_1 and s_2 the results are conveniently tabulated in most textbooks of quantum mechanics. For our own purposes we shall restrict the discussion to two special cases.

Singlet state of two spin-½ particles

Here $s = 0$, and the only value for m is 0. The resulting state vector is

$$|\Psi_{\text{singlet}}\rangle = \frac{1}{\sqrt{2}}(|\alpha(1)\rangle|\beta(2)\rangle - |\beta(1)\rangle|\alpha(2)\rangle) \qquad (102)$$

In this equation the arguments of the spin state vectors are used to distinguish the two particles. So $|\alpha(1)\rangle$ is the eigenket $|\sigma_{1z} = +1\rangle$, and $|\beta(1)\rangle$ is the eigenket $|\sigma_{1z} = -1\rangle$. Similarly, $|\alpha(2)\rangle = |\sigma_{2z} = +1\rangle$ and $|\beta(2)\rangle = |\sigma_{2z} = -1\rangle$.

Singlet states of total spin have the property of rotational invariance. This means that if we act on $|\Psi_{\text{singlet}}\rangle$ with the tensor product of the rotation operators referring to the separate particles, for any given arbitrary rotation, the result will be to reproduce $|\Psi_{\text{singlet}}\rangle$ to within a possible phase factor. For example, consider the rotation through 90° about the positive Y-axis discussed in Section 1.7 (cf. Eq. (85) with $\theta = \pi/2$). This changes $|\alpha\rangle$ into $|\gamma\rangle$ and $|\beta\rangle$ into $|\delta\rangle$. So $|\Psi_{\text{singlet}}\rangle$ is changed to

$$\frac{1}{\sqrt{2}}(|\gamma(1)\rangle|\delta(2)\rangle - |\delta(1)\rangle|\gamma(2)\rangle)$$

But on substituting from Eq. (83), this is seen to be equal again to $|\Psi_{\text{singlet}}\rangle$—i.e. $|\Psi_{\text{singlet}}\rangle$ can be written in the form (102) or equally in the alternative form

$$|\Psi_{\text{singlet}}\rangle = \frac{1}{\sqrt{2}}(|\gamma(1)\rangle|\delta(2)\rangle - |\delta(1)\rangle|\gamma(2)\rangle) \qquad (103)$$

These alternative ways of expanding $|\Psi_{\text{singlet}}\rangle$ will be used in the discussion of the EPR paradox in Chapter 3.

Another problem we shall meet in Chapter 4 is the evaluation of the correlation function

$$c(\mathbf{a}, \mathbf{b}) = \langle(\boldsymbol{\sigma}_1\cdot\mathbf{a})\otimes(\boldsymbol{\sigma}_2\cdot\mathbf{b})\rangle_{|\Psi_{\text{singlet}}\rangle} \qquad (104)$$

Taking the Z-axis along the direction of the unit vector $\mathbf{a}$, and choosing the X-axis so that the unit vector $\mathbf{b}$ lies in the XZ plane, making an angle θ_{ab} with $\mathbf{a}$, as in Fig. 7, we write

$$c(\mathbf{a}, \mathbf{b}) = \langle\Psi_{\text{singlet}}|(\sigma_{1z}\otimes(\sigma_{2z}\cos\theta_{ab} + \sigma_{2x}\sin\theta_{ab})|\Psi_{\text{singlet}}\rangle$$

Using the form (102) for $|\Psi_{\text{singlet}}\rangle$ we can evaluate the action of the Pauli spin operators on $|\Psi_{\text{singlet}}\rangle$ from Eqs. (73) and (74). Hence we obtain the simple result

$$c(\mathbf{a}, \mathbf{b}) = -\cos\theta_{ab} \qquad (105)$$

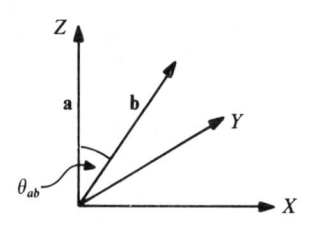

Fig. 7. Choice of coordinate system for evaluating the correlation function $c(\mathbf{a}, \mathbf{b})$ in Eq. (104).

Finally we can evaluate

$$\text{Prob}(\sigma_1 \cdot \mathbf{a} = +1/\sigma_2 \cdot \mathbf{b} = +1)$$

$$= \frac{\text{Prob}(+1, +1)_{\sigma_1 \cdot \mathbf{a}, \sigma_2 \cdot \mathbf{b}}}{\text{Prob}(+1)_{\sigma_2 \cdot \mathbf{b}}}$$

$$= \frac{|\langle \sigma_1 \cdot \mathbf{a} = +1 | \langle \sigma_2 \cdot \mathbf{b} = +1 | (|\Psi_{\text{singlet}}\rangle)|^2}{\|\langle \sigma_2 \cdot \mathbf{b} = +1 | (|\Psi_{\text{singlet}}\rangle)\|^2}$$

$$= \frac{\tfrac{1}{2} \sin^2 \tfrac{1}{2} \theta_{ab}}{\tfrac{1}{2}} = \sin^2 \tfrac{1}{2} \theta_{ab} \qquad (106)$$

where we have employed Eq. (82). This result will also be used later (Section 4.4).

Singlet state of two spin-1 particles

The result in this case is

$$|\Psi_{\text{singlet}}\rangle = \frac{1}{\sqrt{3}} (|S_{1z} = +1\rangle |S_{2z} = -1\rangle + |S_{1z} = -1\rangle$$

$$|S_{2z} = +1\rangle - |S_{1z} = 0\rangle |S_{2z} = 0\rangle) \qquad (107)$$

Again the result is rotationally invariant and can be similarly expressed when spin components of $\mathbf{S}_1$ and $\mathbf{S}_2$ along the Z-axis are replaced by spin components of $\mathbf{S}_1$ and $\mathbf{S}_2$ along an arbitrary direction **n**. Moreover the expansion (107) can be re-expressed in terms of the eigenkets of the spin Hamiltonians H_{s1} and H_{s2} associated with the two particles, by using Eq. (95). The result is

$$|\Psi_{\text{singlet}}\rangle = \frac{1}{\sqrt{3}} (|a+c\rangle |a+c\rangle - |b+c\rangle |b+c\rangle - |a+b\rangle |a+b\rangle)$$

$$(108)$$

where again we use the convention that the left-hand member of a tensor product refers to particle 1 and the right-hand member to particle 2.

This equation will be used in Chapter 6 below.

Notes and References

The classic presentation of Dirac's formulation of QM is his own book on the subject, first published in 1930 and now in its fourth edition, Dirac (1958). The von Neumann approach is presented in von Neumann (1932). A more up-to-date treatment on the same lines is provided by Prugovečki (1981).

For the logico-algebraic approach see Mackey (1963), Jauch (1968) and Varadarajan (1968). Gleason's theorem was first proved in Gleason (1957). Fuller details of the original proof are provided in vol. 1 of Varadarajan (1968). More elementary discussions are given in Krips (1974) and (1977) and Piron (1972). Gleason's paper is reprinted in Hooker (1975), in which many of the original papers on the logico-algebraic approach are conveniently collected.

For more detailed treatment of angular momentum in QM, see standard textbooks such as Schiff (1968) or Messiah (1968). These books really contain more detail than is required for the arguments used in the present work. More elementary texts that can be recommended as essential background to my own book are Gasiorowicz (1974) and Matthews (1974).

2
The Interpretation of Quantum Mechanics

THE preceding chapter presented the formalism of QM together with what we will term the *minimal instrumentalist interpretation* which tells us via the quantization algorithm and the statistical algorithm how the formalism is related to the possible results of measurement and the statistical frequencies with which these measurement results turn up when a measurement is repeated many times (in principle an infinite number of times) on systems prepared in identical QM states.

Already we have introduced two ideas which need considerable discussion. What do we mean by a measurement and what constitutes a state preparation? We shall return to these points in a moment. But first we introduce a *new* sense of interpretation, *different* from that employed in the minimal instrumentalist interpretation (i.e. rules correlating elements of the mathematics with physical quantities). It is simply some account of the nature of the external world and/or our epistemological relation to it that serves to *explain* how it is that the statistical regularities predicted by the formalism with the minimal instrumentalist interpretation come out the way they do. Of course we could simply accept the regularities as 'brute facts' and leave it at that. Interpretations that go beyond the 'brute facts' are idle metaphysical baggage that the austere physicist can and should dispense with. That is to go the way of the positivist, the instrumentalist, the anti-realist. Theories in physics *are* just devices for expressing regularities among observations. Convenience, economy, simplicity are the only criteria for a good theory. The idea of a theory being true in any sense that goes beyond the level of observation is simply not entertained. Indeed, we shall often refer to the formalism of QM plus the minimal instrumentalist interpretation *in the first sense* as the minimal instrumentalist interpretation *in the second sense*. It is no part of the purpose of this book to rehearse the arguments for and against such an attitude to theoretical physics. Suffice it to point to the naive distinction between observation and theoretical terms, and the inability of such an account to make credible the successful novel predictions in which major theoretical advances such as QM abound.

We shall assume for the purposes of this book that theories that lack an interpretation, in the second sense we have indicated, simply do not contribute to our *understanding* of the natural world. Of course this demand for explanation and understanding must not expect too much. If X explains Y, then we can always ask: what explains X? If we just have to accept X, then why not leave X out altogether and just accept Y? The full discussion of this question would lead us into a major area of philosophy of science: the nature of explanation. We content ourselves with a few general remarks. Suppose we accept, as a necessary condition that X explains Y, that Y is logically deducible from X. Then what features of X would lead us to feel that cutting off the demand for understanding Y at the level of just accepting X is an advance on just accepting Y as it stands? The key here seems to be the 'unifying' effect of X. A few general principles about the nature of reality expressed in X comprehend a wide variety of seemingly unconnected observational regularities, including Y. It is often argued that X must be 'picturable' if it is to afford the basis of genuine understanding of Y. What this means is that X must have an *analogue model* in some familiar branch of macroscopic classical physics. We shall not insist on any requirement of 'picturability' ourselves, but it certainly plays a role in some approaches to the interpretation of QM.

Let us illustrate the notion of interpretation by asking the question: 'What can one say about the value of an observable, call it Q, in QM when the state of the system is not an eigenstate of Q?' We will distinguish three general sorts of answer that have been given to this question:

View A: Q has a sharp but unknown value.
View B: Q has an unsharp or 'fuzzy' value.
View C: The value of Q is undefined or 'meaningless'.

We will now discuss these three views in turn.

2.1. View A: Hidden Variables

View A says that QM is not really mysterious at all. It is just a glorified statistical mechanics. There exists an objective external world of entities with well-defined properties which are simply discovered by measurement. The existence of precise possessed values for observables in the pre-measurement state serves to explain why we get a precise result on measurement. Nothing is ever unsharp or fuzzy or undefined or meaningless. Of course, we would have to do a lot more work to fill out view A into a fully-fledged interpretation. We would need to specify a complete dynamics of the possessed values of observables, and this would have to incorporate some account of why it is impossible to prepare states in which non-commuting or incompatible observables simultaneously have zero dispersion. This is the programme envisioned in the so-called hidden-variable interpretation of QM.

The idea can be expressed in the following way. We introduce a set of 'hidden' variables which we denote collectively by the symbol λ. For given λ, the values of all observables are specified as the values of appropriate real-valued functions defined over the domain Λ of possible values for the hidden variables. For observable Q, let us denote the value of Q in the QM state $|\psi\rangle$ by the symbol $[Q]^{|\psi\rangle}$. Considered now as a function $[Q]^{|\psi\rangle}: \Lambda \to \mathbb{R}$, we represent the value of Q when the hidden variables have the value λ by $[Q]^{|\psi\rangle}(\lambda)$. Let us denote the probability density function for the hidden variables in the state $|\psi\rangle$ by ρ_ψ. So $\rho_\psi(\lambda)d\lambda$ measures the probability that the collective hidden variable lies in the range λ to $\lambda + d\lambda$. Then the expectation value of Q in the state $|\psi\rangle$ is

$$\langle Q \rangle_\psi = \int_\Lambda [Q]^{|\psi\rangle}(\lambda) \rho_\psi(\lambda) d\lambda \qquad (1)$$

where we extend the integration in (1) over the whole range Λ of values allowed for the hidden variables. The hidden-variable interpretation of QM now requires us to answer affirmatively the following question. Can we choose the functions $[Q]^{|\psi\rangle}$ and the probability densities ρ_ψ for all states $|\psi\rangle$ and all observables Q in such a way that the expectation values $\langle Q \rangle_\psi$ calculated by (1) agree with the QM expectation values given by the statistical algorithm discussed in the previous chapter?

It is important to notice that what we have termed the hidden-variable interpretation has already introduced some additional assumptions over and above the original idea that all observables always possess sharp values. We are regarding the space Λ as a probability space, i.e. as a space of hidden variables equipped with a probability measure, that expresses our *ignorance* in the QM state $|\psi\rangle$ of the exact values of the hidden variables. The observables are now expressed as random variables over this probability space. As a result of this construction all observables, whether compatible or not, have well-defined joint probability distributions. It has been considered by some authors, notably Arthur Fine, that the assumption of joint distributions for incompatible observables is an unreasonable feature to impose.

Returning to Eq. (1), the idea behind the hidden-variable interpretation is that the state description afforded by $|\psi\rangle$ is incomplete and must be supplemented by the specification of λ in order that the values of all observables should be sharply specified. The QM expectation value then arises by averaging the sharp values over the hidden variables, weighted in accordance with the probability density $\rho_\psi(\lambda)$. λ is called a hidden variable because it cannot itself be directly measured or controlled, a limitation which is necessary in order to prevent the *preparation* of states which are dispersion-free for all observables, including ones which are incompatible (this will be explained in Section 2.5 below). Notice that the notation $[Q]^{|\psi\rangle}(\lambda)$ indicates that the value of Q for a given value of λ may depend on the QM state $|\psi\rangle$. In some discussions of hidden-variable theories it is assumed that this is not the case, i.e. that $[Q]^{|\psi\rangle}(\lambda)$ does not depend on $|\psi\rangle$, but only on λ. There is really no loss of generality, since we can always include $|\psi\rangle$ as part of λ, but hidden variable is then a misnomer.

Interpretations of QM that adopt view A are often described as realist. Statements of such an interpretation are true or false in virtue of whether the properties and relations they assert correspond to states of affairs in an objective external world, which exists independently of the experience of human minds. There are two aspects to this idea of a realist interpretation of QM:

1. The interpretation is more than just an instrument for deriving empirical predictions. In this sense it is *anti-instrumentalist*.
2. It is *anti-idealist*, in the sense that human consciousness plays no role in the specification of the interpretation.

One reason for the appeal of view A is that it is closely modelled on the methods and concepts of classical physics. In this sense it could be characterized as reducing the unfamiliar to the familiar. But is such an interpretation of QM possible? This is in fact the principal question with which this book is concerned. We shall see later that, by building a few more features into view *A*, we can actually show that this is not a possible way of interpreting, i.e. understanding, QM. But the crucial point here will be the discussion of what these additional features are which make such a strong claim possible. What we are really going to do in this book is to rule out certain metaphysical packages as possible ways of interpreting QM. That may be negative progress—telling us what will not work, rather than a positive affirmation of what picture of reality QM commits us to. But, nevertheless, it seems to us very significant that broad results of this negative character are possible in the philosophy of physics.

2.2. View B: Propensities and Potentialities

The terminology of unsharp or 'fuzzy' values is prevalent in some elementary textbooks on QM. But what is really intended is that the observable Q does not possess a *value* at all. What the QM system does in reality possess is a *propensity* or *potentiality* to produce various possible results on measurement, in respect of the observable Q. View *B* says that QM *is* mysterious, in that new concepts over and above those employed in classical physics have to be invoked. The locution 'new' is to be understood only in relation to classical physics. In fact, the idea of potentiality was central to Aristotelian physics—crudely, that the acorn 'possessed' the potentiality of becoming an oak-tree, and that all change consisted just of the actualization of potentialities. Heisenberg, in his later writings on the philosophy of QM, was particularly concerned to stress the Aristotelian affinities of this type of interpretation. Another way of expressing view *B* is the concept of 'latent' quantities due to Margenau, which he contrasted with 'possessed' quantities considered in classical physics. Measurement of an observable not in an eigenstate of that observable is supposed to convert latent values into possessed values. This is the general function of measurement on view *B*, to convert unsharp values into sharp values, to actualize potentialities.

An important question must now be considered. Is view *B* a realist interpretation of QM? It has been argued that propensities are not properties of physical systems as such, but rather that they are relational attributes of microsystems plus *repeatable* experimental arrangements or set-ups which serve to manifest the propensities as long-run relative frequencies for the various possible outcomes of measurement. It is sometimes argued that this relational aspect of propensities tells against realism, in the sense that microsystems do not possess properties independently of our experimental probing. But two points should be noticed.

1. While it is true that propensities are manifested, like other dispositional attributes, in the context of experimental arrangements or set-ups, it seems perfectly coherent to regard them as properties of microsystems and truly possessed by them, in advance and independently of how they are manifested. Consider the claim that this windowpane possesses the property of fragility, which will be manifested if a brick is thrown at it. This way of speaking does not suggest that we cannot be thoroughgoing realists about the fragility of windowpanes. So why should we not be realists about the propensities of microsystems? Of course, what is going on with the objectors to this latter claim is that dispositional properties are really secondary qualities—what windowpanes possess are primary qualities of molecular constitution which explain why, when we throw the brick, the windowpane breaks; there is not an extra primary quality of fragility that is also possessed by the windowpane. Now this may be true of windowpanes, but it does not follow that, in microphysics, dispositional properties should not be regarded as irreducibly primary themselves, to pursue the Lockean terminology. There is certainly nothing here to argue for any sort of idealist interpretation of QM on the grounds that all qualities are secondary rather than primary, and a science of purely secondary qualities must lead to Berkelian idealism!

2. This leads to our second point. The experimental probings under which propensities manifest themselves have nothing to do with human minds or consciousness. Everything could work out in a world without human beings at all. The fact that we design experiments and intend them to furnish us with certain sorts of information about the external world is quite irrelevant in this connection.

The conclusion is that view *B* is perfectly consistent with realism, and certainly gives no arguments at all in favour of idealism.

2.3. View C: Complementarity

View *C* is essentially the view taken by Bohr, and is based on the orthodox Copenhagen complementarity interpretation of QM. It is no part of our purpose in this book to give a detailed analysis of

Bohr's views on QM. We will just comment on a few of the salient features:

1. It would be misleading to regard the lack of definability of Q in a non-eigenstate of Q as a reflection of limitations on the measurability of Q—namely, that measurements of Q in such a state produce a spread of results, and are accompanied by an inevitable loss of knowledge of the value of any non-commuting observable Q' whose value is 'sharp' in the QM state in question. By 'sharp' we mean here that the state of the system is an eigenstate of Q', so we can predict (with probability one if not with certainty) what the result of measuring Q' will be. Or, to put the point another way, in preparing a non-eigenstate of Q we inevitably forgo precise knowledge of what value of Q a subsequent measurement of Q would yield. For Bohr the undefinability of Q is not grounded in its unknowability (a positivist type of argumentation); it is rather the other way about. The unknowability of Q (in the sense we have explained) is grounded in its undefinability.

2. The grounds for definability are, in general mutually exclusive, experimental arrangements for displaying QM 'phenomena'.

3. Complementarity is a relationship that exists between mutually exclusive QM phenomena. Although complementary phenomena cannot 'occur' simultaneously, their mutual possibility is necessary for the complete description of quantum-mechanical reality.

4. The description of the experimental arrangement involved in the specification of a QM phenomenon must employ the concepts and language of classical physics.

5. This does not mean, however, that macroscopic apparatus is not itself susceptible to the QM formalism. The question of whether we use quantum mechanics or classical physics is related to the role of the macroscopic apparatus. In so far as it is doing the 'observing', classical physics must be used. However, if the apparatus is itself the object of observation (by other pieces of macroscopic apparatus), then a QM description must be employed.

Broadly, we can say that QM *is* mysterious on view *C*, not in the sense of requiring new concepts but in the sense of recognizing limitations on the applicability of the familiar concepts of classical physics, which may not be definable in certain contexts.

The essential point with Bohr is the primacy of classical concepts for the description of QM phenomena. In a sense, microphysics is reduced to macroscopic classical physics; but this is linked, as we have seen, with the recognition that macroscopic objects are not themselves immune from the formalism of QM.

The difficulty with assessing the complementarity interpretation of QM is undoubtedly the fact that Bohr's own formulation of the general framework of his ideas is vague and ambiguous. From the methodological point of view, the main objection is the finality with which Bohr prohibits even asking certain questions about QM systems. Complementarity was for Bohr a major philosophical discovery, made in the context of quantum physics, but having applications (in Bohr's view) to other areas of human knowledge, such as psychology and sociology. Setting dogmatic limitations on scientific theorizing, on the basis of obscure philosophical preconceptions, is a dangerous prejudice from the standpoint of a conjectural-fallibist approach to the nature of scientific activity. It is for this reason that other approaches to the interpretation of QM are the main business of this book.

2.4. Measurement and State Preparation

In classical physics, the notion of correlative measurement is unproblematic. A correlation is established between the quantity Q possessed by the system S under investigation and some quantity, call it R, characteristic of the measuring apparatus A. So if Q has the value q (in appropriate units), then R has the value $\rho(q)$, where $\rho: \{q\} \to \{r\}$ is an invertible 1:1 map from the set of values $\{q\}$ that can be possessed by Q into the set of values $\{r\}$ that can be possessed by R. The reason for requiring the map to be invertible is simply that different Q-values get correlated with different R-values; so, if we know the value of R, we can uniquely pick out the associated value of Q. So the question of measuring the value of Q reduces to the question of finding the value of R. This seems just to re-set the original question, transferred from S to A. Do we now have to measure R to find out its value by introducing another measurement apparatus A' that correlates R with R', say, and so on? But this regress is avoided by requiring that R is in some sense directly observable, the famous 'pointer readings' that schematize accounts of classical measurements. If Q were itself 'directly' observable, there would be no need to introduce the apparatus A. The function of measurement is to correlate the value of something which is not 'directly' observable with something which is 'directly' observable.

Of course, in order to establish the required correlation between the value of Q on S and R on A requires some sort of interaction to occur between S and A. This interaction may well affect the value of Q. If, before the measurement interaction were established, Q possessed the value q, but after the interaction the value had changed to q', then observing R gives us the value q via the inverse of the function ρ. Measurement is a backward-looking process that is designed to tell us what value Q possessed before the interaction with A was established. But suppose that, knowing q and the details of the interaction with A, we can predict the value q' after the measurement interaction. We could then regard the measurement interaction as a forward-looking device that enables us to specify the value q' possessed by Q after the interaction between S and A has ceased. It is then functioning as a state-preparation device, preparing the state of S to be such that Q has the value q'. If we remeasure Q on S we will discover this value q' and prepare some new value q'' of Q. Of course, the interaction between S and A might be such that the value of Q did not get changed, so $q' = q$. We might call this non-disturbing measurement; but conceptually there are always two distinct aspects of measurement—what it does to A, the true measurement aspect, and what it does to S, what we may call the state-preparation aspect.

Again, in classical physics we may suppose the magnitude of the interaction in principle to be made so small that not only is Q not disturbed, but S remains totally unaffected by the measurement. Let us now see how much of this classical measurement scheme can be retained in the context of QM.

Let us consider the simplest possible case, which we will refer to as *ideal* measurement. Suppose Q is a maximal (nondegenerate) observable with discrete spectrum $\{q_i\}$. Suppose the QM state of S is an eigenstate $|q_i\rangle$ of Q, and let S interact with the apparatus A, also described by the QM formalism, with A initially in the eigenstate $|r_o\rangle$ of the quantity R; and, as a result of the interaction, suppose the state of A changes from $|r_o\rangle$ to $|r_i\rangle$, while the state of S is unchanged at $|q_i\rangle$.

So the state of the joint system $S + A$ goes from $|q_i\rangle|r_o\rangle$ before the interaction to $|q_i\rangle|r_i\rangle$ after the interaction. Considered as observables on the joint system, the operators for Q and R must be written as $Q \otimes I$ and $I \otimes R$ respectively, where the first factor in the tensor product refers to S and the second factor to A. Since the states before and after the interaction are simultaneous eigenstates of $Q \otimes I$ and $I \otimes R$, we can regard the value q_i as possessed by S both before

The Interpretation of Quantum Mechanics

and after the interaction, the value r_o as possessed by A before the measurement, and the value r_i as possessed by A after the measurement. (The conditions under which observables can be said to *possess* values will be examined in much more detail in Chapter 3.) Everything is quite parallel to the classical case of non-disturbing measurement, the value r_i of R being correlated with the value q_i of Q.

But now suppose that the initial state is given by

$$|\psi\rangle = \left(\sum_i c_i |q_i\rangle\right)|r_o\rangle \qquad (2)$$

so the state of S is a superposition of eigenstates of Q with probability amplitudes c_i.

As a result of the linearity of the time-evolution operator in QM, the state of the joint system after measurement will be

$$|\psi'\rangle = \sum_i c_i |q_i\rangle |r_i\rangle \qquad (3)$$

where we suppose all the r_i are distinct.

Let us work in terms of statistical operators. Before the measurement, the statistical operator for the joint system is

$$W = P_{|\psi\rangle} = P_{(\sum_i c_i |q_i\rangle)|r_o\rangle} \qquad (4)$$

After the measurement, the statistical operator for the joint system is

$$W' = P_{|\psi'\rangle} = P_{\sum_i c_i |q_i\rangle |r_i\rangle} \qquad (5)$$

But what we would expect for the statistical operator after the measurement is not W' but W'', where

$$W'' = \sum_i |c_i|^2 P_{|q_i\rangle |r_i\rangle} \qquad (6)$$

W'' is a mixed state describing an ensemble of systems in states $|q_i\rangle|r_i\rangle$, such that the probability of finding the state $|q_i\rangle|r_i\rangle$ in the mixture is $|c_i|^2$. The component states of the mixture are ones in which we can properly speak of Q as possessing the value q_i and R possessing the value r_i; and the probabilities $|c_i|^2$ just reflect our ignorance of which sharp values of R (and of Q) are going to turn up on any particular measurement occasion. This way of interpreting a mixed state in QM is usually referred to as the ignorance interpretation of mixtures. We shall have more to say about it in a moment.

The so-called problem of measurement in QM is that what we want as the statistical operator in the final state is W'', whereas what we get, as a result of the QM formalism, is something different, namely W'.

Let us look a little more closely at the difference between W' and W'''. First, W' describes a *pure* state. It is an idempotent operator. (See Mathematical Appendix, p. 174.) $|\psi'\rangle$ is not an eigenstate of either $Q \otimes I$ or $I \otimes R$. The initial state $|\psi\rangle$, although not an eigenstate of $Q \otimes I$, was an eigenstate of $I \otimes R$. The effect of the measurement interaction has been twofold:

1. It has left the joint system still in a non-eigenstate of $Q \otimes I$.
2. It has changed the joint system into a non-eigenstate of $I \otimes R$.

In other words, not only has the mysterious superposition of eigenstates of $Q \otimes I$ that we started with not been eliminated, but, worse still, the apparatus observable R has become infected with superposition itself!

Before discussing possible ways of dealing with the measurement problem, we may note that not all observables for the joint system $S + A$ can distinguish between the states specified by W' and W''', in the sense of having different expectation values in the two states. Let us call an observable J for the joint system *insensitive* if it is unable to distinguish W' and W'''. For any J, consider the two expectation values

$$\langle J \rangle_{W'} = \langle \psi' | J | \psi' \rangle$$

$$= \sum_i |c_i|^2 \langle J \rangle_{|q_i\rangle|r_i\rangle} + \sum_{i \neq j} \sum_j c_i^* c_j \langle q_i| \langle r_i | J | q_j \rangle | r_j \rangle \quad (7)$$

while

$$\langle J \rangle_{W'''} = \sum_i |c_i|^2 \langle J \rangle_{|q_i\rangle|r_i\rangle} \quad (8)$$

The difference between (7) and (8) lies in the second 'interference' term that contributes to $\langle J \rangle_{W'}$.

The condition for J to be insensitive is clearly that this term should vanish. This will certainly be the case if J is of the form $Q' \otimes I$ where Q' is any observable referring to S, since

$$\langle q_i| \langle r_i | Q' \otimes I | q_j \rangle | r_j \rangle = \langle q_i | \langle r_i | ((Q'|q_j\rangle) \otimes |r_j\rangle)$$
$$= \langle q_i | Q' | q_j \rangle \langle r_i | r_j \rangle$$
$$= \langle q_i | Q' | q_j \rangle \delta_{ij}$$
$$= 0 \text{ for } i \neq j.$$

Moreover, it is easy to see that

$$\langle Q' \otimes I \rangle_{W''} = \langle Q' \rangle_{W^s} \tag{9}$$

where

$$W^s = \sum_i |c_i|^2 P_{|q_i\rangle} \tag{10}$$

This shows that, relative to measurements of any observable pertaining to the system S alone, the state of S, when the joint system is in the state W', is characterized by the statistical operator W^s given in (10). Or, in other words, the pure state W' behaves like a mixture for subsequent measurements performed on S alone.

A similar result holds for any observable R' pertaining to the apparatus alone

$$\langle I \otimes R' \rangle_{W'} = \langle R' \rangle_{W^A} \tag{11}$$

where

$$W^A = \sum_i |c_i|^2 P_{|r_i\rangle} \tag{12}$$

So, to distinguish between W' and W''', we must perform some sort of correlation measurement between observables pertaining to S and observables pertaining to A. It is important to notice, however, that $Q \otimes R'$ for any R' and $Q' \otimes R$ for any Q' are insensitive observables. To check this, note that

$$\langle q_i | \langle r_i | Q \otimes R' | q_j \rangle | r_j \rangle = \langle q_i | \langle r_i | ((Q|q_j\rangle) \otimes (R'|r_j\rangle))$$
$$= q_j \langle q_i | q_j \rangle \langle r_i | R' | r_j \rangle$$
$$= 0 \text{ if } i \neq j$$

Similarly for $Q' \otimes R$.

In order to distinguish between W' and W''' by means of an observable which is the tensor product of observables appropriate to S and A, we must measure the expectation value of an observable like $Q' \otimes R'$, where Q' is not diagonal in the $\{|q_i\rangle\}$ representation and R' is not diagonal in the $\{|r_i\rangle\}$ representation. The crucial point about the measurement problem is that there are observables on the joint system which are not insensitive.

There is an enormous literature in the philosophy of QM discussing the measurement problem. The problem arises because we have tried to describe the measurement interaction and the behaviour of the apparatus as governed by the laws of QM. One 'solution' is to deny this unified approach. There are two sorts of time evolution in

QM, the unitary time evolution governed by the time-dependent Schrödinger equation, and a quite different sort of time evolution, which is appropriate to measurement interactions, and leads from W to W''' instead of from W to W'. The difficulty here is to decide what distinguishes measurement interactions from other sorts of interaction in QM. Other solutions have concentrated on the macroscopic nature of the measuring apparatus and the irreversible character of measurement procedures. As a schematic remark, suppose that the measuring process actually establishes a chain of correlations between successive systems A, A', A'', ... and the original system S, then the final statistical operator would be

$$W' = P_{\sum_i c_i |q_i\rangle |r_i\rangle |s_i\rangle |t_i\rangle \ldots} \qquad (13)$$

where r_i, s_i, t_i ... are eigenvalues of observables R, S, T ..., associated with the systems A, A', A'' ....

The *required* mixture would be

$$W''' = \sum_i |c_i|^2 \, P_{|q_i\rangle |r_i\rangle |s_i\rangle |t_i\rangle \ldots} \qquad (14)$$

and in this case W' and W''' can only be distinguished by multiple correlation measurements of an observable like

$$Q' \otimes R' \otimes S' \otimes T' \ldots$$

where none of the component observables are diagonal in the corresponding representations in which Q, R, S, T ... are diagonal. Clearly, as the number of steps in the 'amplification' chain increases, the practical difficulty of distinguishing W' from W''' by the required multiple correlation experiments becomes more and more delicate, and, in an idealized limit as the number of steps becomes infinite, may even be declared insuperable. But it is fair to say that this aspect of the interpretation of QM remains highly problematic. What we shall do in this book is to assume that the measurement problem has been solved, for example simply to ignore the objections to the 'two time evolutions' solution. We are going to concentrate on other interpretative problems of QM.

Suppose, then, that in some way or other we have established that the effective statistical operator after the measurement interaction is the mixture W''' rather than the pure state W'. Does this establish that the component states of the mixture are really $|q_i\rangle |r_i\rangle$, reverting for the moment to the simple one-step measurement scheme? The

The Interpretation of Quantum Mechanics

question presents itself since W'' does not have a unique resolution in terms of $P_{|q_i\rangle|r_i\rangle}$. Mixtures of other projection operators may have the same statistical operator. Write $|\psi_i\rangle = |q_i\rangle|r_i\rangle$. So $W'' = \sum_i |c_i|^2 P_{|\psi_i\rangle}$. Then, as an extreme but instructive example, suppose that the system S is described by an N-dimensional Hilbert space, so i runs from 1 to N in the summation defining W''', and all the probability weights $|c_i|^2$ are equal, i.e. $|c_i|^2 = 1/N$ for all i. In this case we can show

$$W'' = \frac{1}{N}\sum_{i=1}^{N} P_{|\psi_i\rangle} = \frac{1}{N}\sum_{i=1}^{N} P_{|\phi_i\rangle} \tag{15}$$

where the kets $|\phi_i\rangle$ are derived from the kets $|\psi_i\rangle$ by any unitary transformation.

Thus suppose

$$|\phi_i\rangle = \sum_{j=1}^{N} c_{ij} |\psi_j\rangle$$

where the matrix c_{ij} is unitary so

$$\sum_{i=1}^{N} c_{ij} c_{ik}^* = \delta_{jk}$$

Then the RHS of (15) is

$$\frac{1}{N}\sum_{i=1}^{N} |\phi_i\rangle\langle\phi_i| = \frac{1}{N}\sum_{i=1}^{N}\sum_{j=1}^{N}\sum_{k=1}^{N} c_{ij} c_{ik}^* |\psi_j\rangle\langle\psi_k|$$

$$= \frac{1}{N}\sum_{j=1}^{N}\sum_{k=1}^{N} \left(\sum_{i=1}^{N} c_{ij} c_{ik}^*\right) |\psi_j\rangle\langle\psi_k|$$

$$= \frac{1}{N}\sum_{j=1}^{N}\sum_{k=1}^{N} \delta_{jk} |\psi_j\rangle\langle\psi_k|$$

$$= \frac{1}{N}\sum_{j=1}^{N} |\psi_j\rangle\langle\psi_j| = \frac{1}{N}\sum_{i=1}^{N} P_{|\psi_i\rangle}$$

$$= \text{LHS of (15)}$$

So an equiprobable mixture of states $|\psi_i\rangle$ has exactly the same statistical operator as an equiprobable mixture of states $|\phi_i\rangle$. Hence, if we are just given W'', it is impossible to say which set of states it is a mixture of.

This sort of result, and the generalization of it to the case where the probability weights are unequal (in which case the alternative states in the mixture turn out not to be orthogonal) have led some people to throw doubt on the ignorance interpretation of mixtures. But all that we can really infer from this ambiguity is that we are in fact more ignorant than we thought—we do not know in which state among the set $\{|\psi_i\rangle\}$ the joint system $S+R$ has been left as a result of the measurement, and we do not know even if the choice of possible states should be restricted to $\{|\psi_i\rangle\}$. At all events, if the $\{|\psi_i\rangle\}$ are states in which some directly observable feature R of the apparatus A is diagonal, the statistical operator W''' is *consistent* with the claim that the result of measurement is to leave A in a state which is a mixture of states in which R has a variety of sharp values r_i. It is perhaps unsatisfactory that we cannot make the *stronger* statement that the result of measurement actually is to leave A in a state where R has a sharp but unknown value.

Let us move on from this defect in the formal theory of measurement, and consider the state-preparation aspect of measurement. Assuming that we have obtained the mixture W''' by means of a physical interaction between S and A, we can now select a sub-ensemble of $S+A$ systems for which R has some particular value r_j. This sub-ensemble will be described by the statistical operator

$$W'''' = P_{|q_j\rangle|r_j\rangle} \tag{16}$$

W'''' is a pure state in respect of both Q and R. It describes S systems with the sharp value q_j for Q. Notice that W'''' is not produced as a result of physical interaction from W'''. It is really a 'mental' operation that we perform when we decide to focus our attention on the sub-ensemble of $S+A$ systems which show the value r_j for the directly observable feature R of A. That is not to say that we cannot arrange to collect or channel all S systems that, considered as $S+A$ systems, show R having the value r_j and hence Q having the value q_j, into some physical space where they can be used for subsequent experimentation. But the decision not to include any other S systems in the specification of the statistical operator is not a physical but a mental operation.

This discussion has relevance to the so-called Projection Postulate in the quantum theory of measurement. It is often stated that, if Q on S is measured and found to have the value q_j, then the state vector for S directly after the measurement is the eigenket $|q_j\rangle$, or equivalently the statistical operator is $P_{|q_j\rangle}$. This is certainly true for the ideal measurement scheme we have been considering (essentially specified by Eq. (3)), provided that we consider not just the nonselective measurement interaction leading to the statistical operator W'' but the selective step leading from W'' to W'''. However it is much more realistic to consider non-ideal measurement schemes in which the state $|q_i\rangle|r_0\rangle$ for the joint system $S+A$ before the measurement interaction goes to $|u_i\rangle|r_i\rangle$ after the measurement interaction where the state $|u_i\rangle$ is unequal to $|q_i\rangle$, although of course it can be expanded in terms of the set $\{|q_i\rangle\}$. By the same analysis as we used above it is easy to see that if the initial state of S is a superposition, then the selective stage of measurement leads to the statistical operator $P_{|u_j\rangle}$, if selected according to the observed value r_j for R. So in this situation the Projection Postulate is simply false. Of course, the special feature of the ideal measurement case is that, if we measure Q on S and find the result q_j, then a subsequent measurement on this particular system will give the value q_j again (at any rate there is probability one that this will happen). In a sense we can say that, for ideal measurement, the distinction between state preparation and measurement is unnecessary. Ideal measurement just prepares the eigenket of the observable which is being measured. But this is certainly not an inevitable concomitant of measurement interactions in their role of state preparation.

2.5. The Uncertainty Relations

Let us pursue the question of the effect of measurement on the microsystem S, the state-preparation aspect of measurement. In the ideal measurement scheme outlined in the preceding section, the effect of the measurement interaction produces no change on the statistics associated with subsequent measurements of the observable Q itself. For example, before the measurement interaction the expectation value of Q in the state $W^S_{\text{initial}} = P_{\sum_i c_i |q_i\rangle}$ is simply $\sum_i |c_i|^2 q_i$, and this is the same as the expectation value of Q in the state $W^S = \sum_i |c_i|^2 P_{|q_i\rangle}$, specified in (10) as the state appropriate to measurements of observables pertaining to S after the measurement interaction. We shall write this henceforth as W^S_{final}, to emphasize that this is the state after the measurement interaction.

The fact that the statistics of Q does not change in going from W^S_{initial} to W^S_{final} expresses the fact that the ideal measurement scheme is non-disturbing so far as the observable Q is concerned. But this does not mean that the measurement of Q may not 'disturb' S so far as other observables Q' are concerned.

Thus, in general

$$\langle Q' \rangle_{W^S_{\text{initial}}} = \sum_i |c_i|^2 \langle Q' \rangle_{|q_i\rangle} + \sum_{i \neq j} \sum_j c_i^* c_j \langle q_i | Q' | q_j \rangle \qquad (17)$$

and

$$\langle Q' \rangle_{W^S_{\text{final}}} = \sum_i |c_i|^2 \langle Q' \rangle_{|q_i\rangle} \qquad (18)$$

If Q' is not itself diagonal in the representation $\{|q_i\rangle\}$, (17) and (18) will in general differ, due to the additional 'interference' terms in (17) which are absent from (18). The disturbing effect of a measurement of Q on observables Q' other than Q is often thought to be clarified by consideration of the famous uncertainty relations in QM. We begin by stating the result formally. Let A and B be any two observables pertaining to a system S in an arbitary state $|\psi\rangle$.

Define $\quad \Delta A = (\langle (A - \langle A \rangle_{|\psi\rangle} \cdot I)^2 \rangle_{|\psi\rangle})^{\frac{1}{2}}$

and $\quad \Delta B = (\langle (B - \langle B \rangle_{|\psi\rangle} \cdot I)^2 \rangle_{|\psi\rangle})^{\frac{1}{2}} \qquad (19)$

Define the Hermitian operator C by the relation

$$[A, B] = AB - BA = iC \qquad (20)$$

Then the uncertainty relations state that

$$\Delta A \cdot \Delta B \geq \tfrac{1}{2} |\langle C \rangle_{|\psi\rangle}| \qquad (21)$$

The proof is standard:

Define $\quad T = A - \langle A \rangle_{|\psi\rangle} \cdot I + i\rho(B - \langle B \rangle_{|\psi\rangle} \cdot I) \qquad (22)$

for any real number ρ.

Then $\quad \langle T^\dagger T \rangle_{|\psi\rangle} \geq 0 \qquad (23)$

This follows since $\langle T^\dagger T \rangle_{|\psi\rangle} = \langle \psi | T^\dagger \cdot T | \psi \rangle$ is the square of the norm of the vector $T|\psi\rangle$, and hence cannot be a negative quantity.

But

$$\langle T^\dagger T \rangle_{|\psi\rangle} = (\Delta A)^2 - \rho \langle C \rangle_{|\psi\rangle} + \rho^2 (\Delta B)^2 \qquad (24)$$

The RHS of Eq. (24) is thus a quadratic form in ρ which is non-negative. This implies that the discriminant

$$(\langle C \rangle_{|\psi\rangle})^2 - 4(\Delta A)^2 \cdot (\Delta B)^2 \leq 0$$

or
$$\Delta A \cdot \Delta B \geq \tfrac{1}{2} |\langle C \rangle_{|\psi\rangle}|$$

as stated in (21).

The interpretation of the result (21) is as follows. If we prepare a system S in a pure state $|\psi\rangle$, then repeated measurements of A in an ensemble of identically prepared such systems will yield a standard deviation ΔA around the mean value $\langle A \rangle_{|\psi\rangle}$. Similarly, measurements of B will yield a standard deviation ΔB.

Eq. (21) demonstrates a reciprocal limitation on how small the dispersions ΔA, ΔB in A-measurements and B-measurements can be made in the QM state $|\psi\rangle$, in terms of the expectation value in the state $|\psi\rangle$ of the commutator of the observables A and B. If $\langle C \rangle_{|\psi\rangle}$ does not vanish, then the dispersion in A-measurement results cannot be less than $\tfrac{1}{2} |\langle C \rangle_{|\psi\rangle}|/\Delta B$, and similarly ΔB cannot be less than $\tfrac{1}{2}|\langle C \rangle_{|\psi\rangle}|/\Delta A$.

The following points should be noticed:

1. If $|\psi\rangle$ is an eigenstate of A or B, then $\langle C \rangle_{|\psi\rangle}$ always vanishes.
2. In a finite-dimensional Hilbert space, or in an infinite-dimensional Hilbert space where either A or B has a purely discrete spectrum, C cannot be a constant multiple of the identity. This is because, in a representation in which A or B are diagonal, the diagonal elements of C are all zero, in accordance with remark 1, while the identity operator has unit diagonal elements in any representation.

The most famous application of the uncertainty relations is to the observables P and Q representing momentum and position. Considering the one-dimensional case we have the relation

$$[Q, P] = i\hbar \cdot I \tag{25}$$

So C is just $\hbar \cdot I$. We learn immediately from this result that position and momentum observables cannot be comprehended in the framework of finite-dimensional Hilbert spaces, and moreover must involve the niceties of continuous spectra (see remark 2 above). But the uncertainty relations are still applicable and read simply

$$\Delta Q \cdot \Delta P \geq \tfrac{1}{2}\hbar \tag{26}$$

Let us return to the disturbing effect of Q-measurements on observables Q' that are not diagonal in the $\{|q_i\rangle\}$ representation.

Suppose the initial state $\sum_i c_i |q_i\rangle$ is an eigenstate of Q', so initially $\Delta Q' = 0$. The following argument is often given. When we measure Q, then the final state is a mixture of eigenstates of Q. In any such eigenstate, Q' must exhibit dispersion as a result of the uncertainty relation involving Q and Q'. The effect of the uncertainty relation is to force dispersion in Q' when Q is measured, when before Q was measured Q' exhibited no dispersion. However, for the measurement of quantities with a discrete spectrum, the result that, in an eigenstate of Q, Q' will exhibit dispersion is in general true (unless we are dealing with the special case of a simultaneous eigenstate of Q and Q'—remember that it may be possible for non-commuting operators to have *some* eigenvectors in common). But this cannot be deduced from the uncertainty relations, since, as we have seen in remark 1 above, the RHS of the uncertainty relation vanishes in eigenstates of either observable.

Nevertheless, the uncertainty relations do represent a very significant limitation on our ability, according to the QM formalism, to prepare states which are dispersion-free in respect of *all* possible observables. It is of course easy to prove, from the statistical algorithm, that no QM state can be dispersion-free for all observables. The condition for this would be

$$(\text{Tr } WQ)^2 = \text{Tr}(WQ^2), \quad \forall Q \qquad (27)$$

Apply this condition to $Q = P_{|\psi\rangle}$ for any state $|\psi\rangle$. Then we require

$$(\text{Tr}(WP_{|\psi\rangle}))^2 = \text{Tr}(WP_{|\psi\rangle}^2)$$
$$= \text{Tr}(WP_{|\psi\rangle})$$

Whence

$$\text{Tr}(WP_{|\psi\rangle}) = \langle\psi|W|\psi\rangle = 0 \text{ or } 1, \quad \forall|\psi\rangle$$

The first possibility requires $W = 0$ and the second requires $W = I$, neither of which is consistent with $\text{Tr } W = 1$ (assuming dimension $H > 1$). Notice that the possibility that $\langle\psi|W|\psi\rangle = 0$ for some $|\psi\rangle$ and $= 1$ for other $|\psi\rangle$ is excluded by the plausible fact that $\langle\psi|W|\psi\rangle$ changes in a continuous way as the ket $|\psi\rangle$ changes its direction in Hilbert space. (In this book we shall not develop the machinery to make this fact precise or to show that it is true.)

An important question now arises. Granted that there are limitations on preparing states with simultaneous sharp values for non-commuting observables, does this imply that we cannot simul-

The Interpretation of Quantum Mechanics

taneously measure such observables in states subject to these limitations? Consider the following situation. Suppose at time t_1 we prepare an eigenket of Q (assumed maximal) corresponding to the eigenvalue q, for example by performing an ideal measurement of Q and selecting those systems for which the measurement result shows the value q. So our initial state is $|q\rangle$. Now, at a later time t_2, measure a non-commuting (incompatible) observable Q' (for which $|q\rangle$ is not an eigenket) and suppose we find the value q' for Q'. Suppose the interval $t_2 - t_1$ is so small that we can neglect any time evolution of the state $|q\rangle$ prepared at t_1, i.e. that we regard the state just before the Q'-measurement at t_2 as still effectively $|q\rangle$. Then we might now argue that we have effectively established a situation in which we can claim that we have simultaneously measured Q and Q' at time t_2, the results of the double measurement being q and q'. Of course we only actually measured Q' at t_2, but we could argue counterfactually that, if at t_2 a measurement of Q had been performed instead of Q', then the result would have been q. In this sense we can say that at t_2 we know both what measuring Q' did actually show, viz. q', and also what measuring Q would have shown, viz. q. This argument circumvents the fact that the experimental set-ups for measuring Q and Q' may be incompatible—or complementary, as Bohr would say.

An important point should be noticed here. The time at which simultaneous measurement information can be claimed is just before t_2. This is consistent with the fact that measurement is a backward-looking exercise. Of course, if the measurement of Q' were ideal, then the state of the system *after* t_2 is $|q'\rangle$, and thus, as a result of the Q' measurement, we have lost the ability to predict what the result of a Q-measurement would be.

But now suppose we start with an initial state $|\psi\rangle$ which is not an eigenket of Q, or of Q'. Is there any sense in which we can claim to be able to measure Q and Q' simultaneously in such a state? Schemes for answering this question in the affirmative that have been proposed in the literature involve essentially the following device. Consider a nondegenerate quantity U with eigenvalues $\{u_i\}$, which is compatible (i.e. commutes) with Q', and which possesses the property that in the state $|\psi\rangle$

$$\text{Prob }(q_i)_Q^{|\psi\rangle} = \text{Prob }(u_i)_U^{|\psi\rangle} \tag{28}$$

So we have that U and Q are matched probabilistically in the state $|\psi\rangle$. Let us suppose further that the eigenvalues u_i and q_i are identically equal.

So (28) can be replaced by

$$\overline{\text{Prob}}\ (q_i)_Q^{|\psi\rangle} = \overline{\text{Prob}}\ (q_i)_U^{|\psi\rangle} \tag{29}$$

Then it is claimed that measuring U is the same as measuring Q, since the eigenvalues are equal and the probability for obtaining the eigenvalues are equal (all for the particular state $|\psi\rangle$, remember).

But, since U is compatible with Q' by hypothesis, all we now have to do is to measure U and Q' and record the values obtained, q and q'. Then these two numbers constitute the required joint measurement of Q and Q' in the state $|\psi\rangle$. Before commenting on this scheme, let us give the paradigm example, viz. the so-called 'time of flight' method for simultaneous measurement of position X and momentum P of a particle of mass m.

Suppose at time t_0 we prepare a localized wave-packet in coordinate space (assumed one-dimensional for simplicity) centred on the origin. In accordance with the uncertainty relations, this involves a spread in momentum. Thus consider the normalized Gaussian wave-packet with wave function at t_0

$$\psi(x, t_0) = \sigma^{-1/2}\, \pi^{-1/4}\, e^{-x^2/2\sigma^2} \tag{30}$$

Then the probability density for finding a value of X in the neighbourhood of the value x is

$$\overline{\text{Prob}}\ (x)_X^{|\psi\rangle} = |\psi(x, t_0)|^2$$
$$= \sigma^{-1}\, \pi^{-1/2}\, e^{-x^2/\sigma^2} \tag{31}$$

(We use the notation $\overline{\text{Prob}}$ to indicate a probability density, the appropriate notion for dealing with operators such as X with a continuous spectrum.)

Using Eq. (1.23) the wave function in momentum space is given by

$$\psi(p, t_0) = \frac{1}{\sqrt{2\pi}} \int_{-\infty}^{\infty} e^{-ipx} \psi(x, t_0)\, dx$$
$$= \sigma^{1/2}\, \pi^{-1/4}\, e^{-p^2\sigma^2/2} \tag{32}$$

and

$$\overline{\text{Prob}}\ (p)_P^{|\psi\rangle} = |\psi(p, t_0)|^2$$
$$= \sigma\pi^{-1/2}\, e^{-p^2\sigma^2} \tag{33}$$

The Interpretation of Quantum Mechanics

We notice in passing that, for the state $|\psi_0\rangle$, $\Delta X = \dfrac{\sigma}{\sqrt{2}}$ and $\Delta P = \dfrac{1}{\sqrt{2}\sigma}$.

So the uncertainty product is

$$\Delta X \cdot \Delta P = \tfrac{1}{2}$$

and the uncertainty inequality (26) is satisfied with the equality sign (remember that in Eq. (32) we have assumed $\hbar = 1$). For this reason, the Gaussian wave-packet has the property of being the 'minimum uncertainty' wave-packet.

Now consider the time evolution of the wave-packet (30) at some later time t. This can be readily computed from Eq. (1.25) with the free-particle Hamiltonian

$$H = p^2/2m \tag{34}$$

We have the simple result

$$\psi(p, t) = e^{-ip^2/2m \cdot (t - t_0)} \cdot \psi(p, t_0) \tag{35}$$

(where we have set $\hbar = 1$ in Eq. (1.25)).

This shows that

$$\overline{\text{Prob}\,(p)}_P^{|\psi_t\rangle} = \sigma \pi^{-1/2} e^{-p^2\sigma^2} \tag{36}$$

i.e. the probability density for P-measurements does not change with time.

We now compute $\psi(x, t)$ by inverting the Fourier transform in (1.23). Thus

$$\psi(x, t) = \frac{1}{\sqrt{2\pi}} \int_{-\infty}^{\infty} e^{ipx} \cdot \psi(p, t) \, dp$$

$$= \frac{1}{\sqrt{2\pi}} \int_{-\infty}^{\infty} e^{i(px - p^2/2m \cdot (t - t_0))} \cdot \sigma^{1/2} \pi^{-1/4} e^{-p^2\sigma^2/2} \, dp \tag{37}$$

The integral in (37) can readily be evaluated using the standard formula

$$\int_{-\infty}^{\infty} e^{-a\zeta^2 + 2b\zeta} \, d\zeta = \sqrt{\frac{\pi}{a}} \cdot e^{b^2/a} \tag{38}$$

where $\operatorname{Re} a > 0$.

The result is

$$\psi(x,t) = \sigma^{1/2} \pi^{-1/4} \cdot \frac{1}{\left(\sigma^2 + \frac{i(t-t_0)}{m}\right)^{\frac{1}{2}}} \cdot e^{-\frac{x^2}{2\left(\sigma^2 + \frac{i(t-t_0)}{m}\right)}}$$

and hence

$$\overline{\text{Prob}}(x)_X^{|\psi_t\rangle} = \frac{\sigma \pi^{-\frac{1}{2}}}{\left(\sigma^4 + \frac{(t-t_0)^2}{m^2}\right)^{\frac{1}{2}}} \cdot e^{-\left(\frac{x^2 \sigma^2}{\sigma^4 + \frac{(t-t_0)^2}{m^2}}\right)} \tag{39}$$

Now consider the observable

$$U = \lim_{t \to \infty} \frac{mX}{t - t_0} \tag{40}$$

So $\overline{\text{Prob}}(u)_U^{|\psi_\infty\rangle} = \frac{dx}{du} \cdot \overline{\text{Prob}}(x)_X^{|\psi_\infty\rangle}$

$$= \lim_{t \to \infty} \frac{t - t_0}{m} \cdot \overline{\text{Prob}}(x)_X^{|\psi_t\rangle}$$

$$= \sigma \pi^{-\frac{1}{2}} e^{-u^2 \sigma^2}$$

Thus, in the limit as $t \to \infty$, P and U have matching probability densities, so a simultaneous measurement of X and U at $t \to \infty$ would, according to the argument under discussion, constitute a simultaneous measurement of X and P. But the simultaneous measurement of X and U consists just in measuring X, since U is simply computed from X by means of (40). But does this constitute a genuine joint measurement of X and P? The fact that P and U have matched probability densities does not allow us to say that, on a particular occasion, if we had measured P directly—for example by deflection in a magnetic field—we would have found the same value as we actually found for U. It is only if that counterfactual could be sustained that a genuine claim to joint measurement of X and P could be made.

Notice the contrast with the first sort of example we discussed, which also involved a counterfactual statement, but one which could be justified.

The fact that the operator identity (40) looks just like the classical connection between the momentum of a particle released in the neighbourhood of the origin and its location at some time in the distant future should not lead us to infer that measuring U really is a measurement of P.

Notice that the reason for taking the limit $\tau \to \infty$ in the above discussion is to allow the initial uncertainty in the position of the particle at time t_0 to get 'swamped' by the spreading of the wave-packet with time.

We conclude this section by commenting briefly on the so-called *Disturbance Theory* of the uncertainty relations. In Heisenberg's original discussion of the uncertainty relations, a crucial part was played by the famous γ-ray microscope example, in which the uncertainty relation for a material particle was 'deduced' from the fact that the more accurately the location of a particle was measured (by using a microscope with very short wavelength 'light' (γ-rays) so as to enhance the resolving power of the microscope), the more *disturbed* was the momentum of the observed particle, in a random and unsurveyable fashion, by Compton scattering between the photons in the light-beam and the material particle.

In effect, the argument consists in showing that a classically described particle gets 'infected' with the QM uncertainty relations when it interacts in a measurement situation with a quantal agent, viz. light described by the QM photon concept.

Let us take a simple example of this type of argument. Consider two particles of equal mass m in one-dimensional motion, with momenta P_1 and P_2, which collide elastically and emerge with momenta P'_1 and P'_2 respectively. We suppose that particle 1 is the quantal agent satisfying an uncertainty relation

$$\Delta X_1 \cdot \Delta P_1 \sim \hbar/2 \qquad (41)$$

while particle 2 has well-defined position and momentum, and we apply conservation of momentum and energy in the collision to give the equations

$$P'_1 + P'_2 = P_1 + P_2 \qquad (42)$$

$$P'^2_1 + P'^2_2 = P^2_1 + P^2_2 \qquad (43)$$

From (42) and (43) we find at once, neglecting the trivial possibility that no collision has occurred,

$$P'_2 = P_1 \qquad (44)$$

Considered as a state-preparation device for particle 2, we have from

(44)

$$\Delta P'_2 = \Delta P_1 \tag{45}$$

Let us specialize to the case where particle 2 is initially at rest, i.e. $P_2 = 0$. Then the instant at which the collision takes place is uncertain by an amount

$$T = \frac{\Delta X_1 m}{P_1} \tag{46}$$

This leads to an uncertainty in the position of particle 2 after the collision given by

$$\Delta X'_2 = T \cdot \frac{P'_2}{m} = \Delta X_1 \cdot \frac{m}{P_1} \cdot \frac{P'_2}{m} = \Delta X_1 \tag{47}$$

From (45) and (47)

$$\Delta X'_2 \cdot \Delta P'_2 = \Delta X_1 \cdot \Delta P_1 \sim \hbar/2 \tag{48}$$

So, as a result of the collision with the quantal agent, the state of the particle 2 has become infected with the uncertainty relations, as shown by (48).

What are we to make of such an argument? It seems pretty incoherent. We start by assuming particle 1 is quantal and particle 2 is nonquantal before the collision. We then give a quasi-classical description of the collision process, and end up by showing that particle 2 is quantal. But a little further analysis will show that particle 1 has been rendered nonquantal! To see this, note that using Eqs. (42) and (43) it follows also that

$$P'_1 = P_2 = 0 \tag{49}$$

So, whatever the initial uncertainty in P_1, the final momentum of particle 1 is fixed at the definite value 0 and

$$\Delta P'_1 = 0 \tag{50}$$

We have also

$$\Delta X'_1 = T \cdot \frac{P'_1}{m} = 0 \tag{51}$$

From (50) and (51)

$$\Delta X'_1 \cdot \Delta P'_1 = 0 \tag{52}$$

Indeed, with this sort of argumentation it is possible to prove *and* disprove the uncertainty relations (see references below).

The correct way to understand the uncertainty relations is to see that they represent an inherent limitation on the sort of states which can be produced for QM systems, but that they cannot be 'explained' or 'deduced' by the naive disturbance argument.

Notes and References

The best general references for the philosophy of QM are D'Espagnat (1976) and the monumental work by Jammer (1974). The reader should be aware that the word 'interpretation' is used with a number of different senses in philosophy of science. In this book we keep consistently to the usages explained in the text. For a clear discussion, see chapter 1 of Jammer's book. There is an enormous literature on the general topic of explanation in science. The relation between explanation and unification is explored in Kitcher (1981). The role of models in physics is discussed in Redhead (1980). Hidden variables in QM are given a comprehensive treatment in Belinfante (1973). Propensities were introduced in the philosophy of probability by Popper (1957). See also Popper (1982a), (1982b), and (1983). For a defence of the view that propensities can be regarded as attributes of QM systems rather than experimental set-ups, see Mellor (1971), chapter 4. The idea of latent quantities was introduced by Margenau (1954). For a general philosophical discussion of latency, see McKnight (1958).

For Bohr's philosophy of QM and the notion of complementarity, the main primary references are the three collections of essays, Bohr (1934), (1958), and (1963). The best attempt at a coherent exegesis of Bohr's writings is Scheibe (1973) chapter 1. See also Folse (1985) and Redhead (1987). For the general theory of measurement in classical physics the best introduction is Ellis (1966). The scheme of ideal measurement and the associated Projection Postulate are due to von Neumann (1932). Von Neumann showed that formally ideal measurement is always possible. The fact that, in practice, ideal measurement may be the exception rather than the rule was particularly emphasized by the result of Wigner (1952) that ideal measurement is only possible for quantities which commute with all additive conserved quantities in the measurement interaction. See Araki and Yanase (1960) and Yanase (1961) for further discussion of this point. The ignorance interpretation of mixtures is further discussed by van Fraassen (1972) and Park (1973). A rather general proof of the insolubility of the measurement problem within the framework of unitary time evolution has been provided by Fine (1970). For a critique of this and other insolubility proofs in the literature, see Brown (1986).

The uncertainty relations and the various associated thought experiments are expounded in the classic work of Heisenberg (1930). For the proof of the uncertainty relations from the formalism of QM, see also Robertson (1929). The proof given in the text that there are no dispersion-free states in QM follows von Neumann (1932).

Joint measurability of non-commuting observables is discussed by Popper (1934), Popper (1967), Arthurs and Kelly (1965), She and Heffner (1966), Park and Margenau (1968), and Fitchard (1979). A detailed critique of the Disturbance Theory of the uncertainty relations is given by Brown and Redhead (1981).

A useful collection of reprints of many of the classic papers in the philosophy of QM can be found in Wheeler and Zurek (1983).

Recent work on Bohr's philosophy includes Murdoch (1987) and Honner (1987).

Important new books concerned with a wide range of issues in the philosophy of QM include Gibbins (1987) and Krips (1987).

3
The Einstein–Podolsky–Rosen Incompleteness Argument

In 1935 Einstein, Podolsky, and Rosen (EPR) produced a famous argument for the incompleteness of the minimal instrumentalist interpretation of QM. In this chapter we shall present a version of the argument, and comment on its significance. We begin with a necessary condition for a theory in physics to be complete.

Necessary Condition for Completeness (P):
Every element of the physical reality must have a counterpart in the physical theory.

This is a straight quotation from EPR (1935).

The idea behind it is very simple. The language in which the physical theory is formulated must be sufficiently rich that every proposition signifying putative relations between the 'elements of reality' can at least be expressed in the language of the theory.

P is not a sufficient condition. A sufficient condition might involve the additional requirement that every proposition concerning the 'elements of reality' is provably true or false in accordance with the axiomatic-deductive structure of the theory (supplemented by the specification of contingent initial conditions). This sort of completeness is the subject of the famous Gödel result in mathematical logic, that every first-order theory (a first-order theory is one which does not allow quantification over predicate variables) which includes a significant fragment of arithmetic is actually incomplete according to this additional requirement. The condition P has nothing to do with these Gödelian complications. It is solely concerned with the expressive power of the language in which the theory is formulated. If a theory is incomplete by the failure of this criterion, it should be noted that this is consistent with the theory asserting the existence of elements of reality which in fact do not exist, not just that it fails to refer to all those elements of reality which do exist.

In order to apply the principle *P* in a proof of the incompleteness of QM, we must have at hand some means of identifying the 'elements of reality'. To show that QM is incomplete, the idea of EPR is simply to demonstrate an element of reality which does *not* have a counterpart in the theory. EPR proceed to introduce a sufficient condition for identifying elements of physical reality. We shall present the condition in a form that is slightly different from EPR (and will comment on the difference later).

Sufficient Condition for Element of Reality (R):
If we can predict with certainty, or at any rate with probability one, the result of measuring a physical quantity at time *t*, then at the time *t* there exists an element of reality corresponding to the physical quantity and having a value equal to the predicted measurement result.

The reality criterion *R* certainly seems eminently reasonable. It really depends on a version of *Inference to the Best Explanation*. The best explanation of why we make the successful measurement prediction at time *t* is that there exists an element of reality at time *t*, having that value and predictable in accordance with the physical theory at issue, which is then simply discovered by the measurement. Arguing contrapositively, if there were no element of reality at time *t* then we should not expect to be able to make a successful prediction of the measurement result at time *t*, since predictability can only be expected to arise as a result of regularities underlying the behaviour of elements of reality. Notice that we are not claiming that all elements of reality have predictable values—a thesis of determinism—only that, if it is predictable, then there must exist an element of reality to explain this possibility.

Suppose I look at a table, turn my back and predict that, if I look again, assuming no outside intervention, then I will see the table again. The predictability of my seeing the table again allows me to infer, in accordance with *R*, that there exists an element of reality corresponding to the table at the instant when I look again. In this case the element of reality exists both at the time t_0 the prediction is effected and the time *t* at which it is effective. But in order to show that an element of reality exists at the time t_0 requires another principle, that will allow us to infer that the element of reality was not brought into being at some time between t_0 and *t*. In the case of the table, this is provided by the assumption that there was no outside interference with the state of the table during the period when my back was turned.

To take another example, suppose I predict at t_0 that at a later instant t' some mechanical contrivance will throw a stone in a pond, and hence that at a still later time t a ripple will be observed to strike the bank, then R allows me to infer that there really is a ripple at time t, but this was brought into existence at time t' when the stone was thrown and did not exist at time t_0 when the original prediction was made.

Let us now return to the situation in QM. We employ the notation already introduced in Section 2.1. Denote by $[Q]$ the value of an element of reality corresponding to the observable Q. We shall sometimes use $[Q]$ to denote the element of reality itself. The meaning should always be clear from the context. Consider the three views introduced in Chapter 2. On view A, $[Q]$ will exist at all times in all quantum states for all observables. But views B and C claimed that in non-eigenstates of Q, $[Q]$ did not exist.

Since $[Q]$ may depend partly on what particular QM state the system happens to be in, we shall use the notation $[Q]^{|\phi\rangle}$ to indicate the value that Q possesses, on a particular occasion, in the state $|\phi\rangle$. Then we may employ R to obtain the following result which we call

The Eigenvector Rule: $$[Q]^{|q_i\rangle} = q_i \qquad (1)$$

where $|q_i\rangle$ as usual denotes the ith eigenket of Q belonging to the eigenvalue q_i.

The Eigenvector Rule shows that, on any interpretation of QM, if we allow R, then in an eigenstate of Q there does always exist an element of reality corresponding to Q, and having a value equal to the associated eigenvalue. (This serves to justify the remarks made below Eq. (2.6) about the component states of a mixture as exhibiting possessed or sharp values of the relevant observables.)

Consider a QM system consisting of two spin-$\frac{1}{2}$ particles, in the singlet state of their total spin, and widely separated spatially, so that there is no significant overlap of the spatial wave functions of the two systems. Such a state can be produced, for example, in low-energy p-p scattering which is well known to proceed via the singlet (antisymmetric) state of the total spin of the two protons (since the S-wave spatial wave-function that dominates low-energy scattering is symmetric under exchange of particles, the Pauli Principle forces the spin part of the state vector to be antisymmetric). From Eq. (1.102) we know that this spin state is given by

$$|\Psi_{singlet}\rangle = \frac{1}{\sqrt{2}}(|\alpha(1)\rangle|\beta(2)\rangle - |\beta(1)\rangle|\alpha(2)\rangle) \quad (2)$$

where

$$\begin{aligned}|\alpha(1)\rangle &= |\sigma_{1z} = +1\rangle \\ |\beta(1)\rangle &= |\sigma_{1z} = -1\rangle \\ |\alpha(2)\rangle &= |\sigma_{2z} = +1\rangle \\ |\beta(2)\rangle &= |\sigma_{2z} = -1\rangle\end{aligned} \quad (3)$$

For the state $|\Psi_{singlet}\rangle$ we can easily compute the conditional probability of finding the measurement result $+1$ for σ_{2z}, given that the measurement result for σ_{1z} is -1. This is given by

$$\begin{aligned}&\text{Prob}(\sigma_{2z} = 1/\sigma_{1z} = -1) \\ &\underset{Df}{=} \frac{\text{Prob}(+1, -1)^{|\Psi_{singlet}\rangle}_{\sigma_{2z}, \sigma_{1z}}}{\text{Prob}(-1)^{|\Psi_{singlet}\rangle}_{\sigma_{1z}}} \\ &= \tfrac{1}{2}/\tfrac{1}{2} = 1\end{aligned} \quad (4)$$

Similarly

$$\text{Prob}(\sigma_{2z} = -1/\sigma_{1z} = +1) = 1 \quad (5)$$

Eqs. (4) and (5) express what are often referred as the mirror-image correlations built into $|\Psi_{singlet}\rangle$. Measuring σ_{1z} enables one to predict that a subsequent measurement of σ_{2z} will show the opposite value to the measurement result for σ_{1z}. Let us put in some times. At t_1 measure σ_{1z}, then we can predict the result of measuring σ_{2z} for any time $t_2 > t_1$ (assuming no interference with the joint system other than the measurement of σ_{1z} at time t_1). Hence we can infer, using R, that $[\sigma_{2z}]$ exists for any time $t_2 > t_1$. Suppose, to be specific, that at time t_1 we performed an ideal measurement on σ_{1z} and obtained the value $+1$. The state of the joint system after this measurement, and selected in accordance with this measurement result, is $|\Psi\rangle = |\alpha(1)\rangle|\beta(2)\rangle$. This is an eigenstate of σ_{2z} with eigenvalue -1, and this corresponds to the fact that at any time $t_2 > t_1$, we can predict the measurement result for σ_{2z} to be -1, and hence infer by R the existence of $[\sigma_{2z}]$ with the value -1. We now want to argue that we can project the existence of $[\sigma_{2z}]$ back to a time $t_3 < t_1$ when the state of the joint system was $|\Psi_{singlet}\rangle$. In order to do this we invoke a locality principle L.

Locality Principle (L):

Elements of reality pertaining to one system cannot be affected by measurements performed 'at a distance' on another system.

The locution 'at a distance' can be understood in two senses, which we distinguish as *Bell locality* and *Einstein locality*. For Bell locality, 'at a distance' means in the absence of causal influences recognized by current physical theories. For Einstein locality, 'at a distance' means at a space-like separation between the space–time locations where the element of reality pertaining to one system exists and the measurement on the other system takes place.

If we accept, provisionally, that special relativity (SR) implies Einstein locality, then we are claiming, in the Einstein version of locality, not just that no known physical causal influence is at work, but that no possible causal influence consistent with the constraints of SR could be effective in inducing the change in the element of reality.

Now apply L to the element of reality $[\sigma_{2z}]$ at the time t_1. This shows that no change in that element of reality can be effected as a result of the measurement of σ_{1z}; in other words, L enables us to project the existence of $[\sigma_{2z}]$ backwards (with the value -1 in our example) to a time $t_3 < t_1$. But since, at time t_3, the state of the system is $|\Psi_{\text{singlet}}\rangle$ which is certainly not an eigenstate of σ_{2z} (or of course of σ_{1z}), what we have demonstrated is the existence of $[\sigma_{2z}]$ in a non-eigenstate of σ_{2z}. So what we have shown is that, for a special choice of Q and $|\psi\rangle$, $[Q]^{|\psi\rangle}$ exists even when $|\psi\rangle$ is not an eigenstate of Q. And now we can employ P to conclude that the minimal instrumentalist interpretation of QM is incomplete, since in non-eigenstates of Q there is nothing in that interpretation that refers to $[Q]$—indeed, in interpretations of QM such as B and C it is denied that $[Q]$ exists at all in such states.

The argument we have given is sketched schematically in Fig. 8.

We can summarize the EPR argument in the form

$$F \wedge L \to \text{Incompleteness} \qquad (6)$$

where F denotes the formalism of QM with the minimal instrumentalist interpretation, and we use the logical symbol '$\wedge$' for conjunction. Note that only a small, but very significant, fragment of F is actually used in the argument, viz. the existence of non-factorizable states for the joint Hilbert space of the two systems with strict mirror-image correlations in the way described.

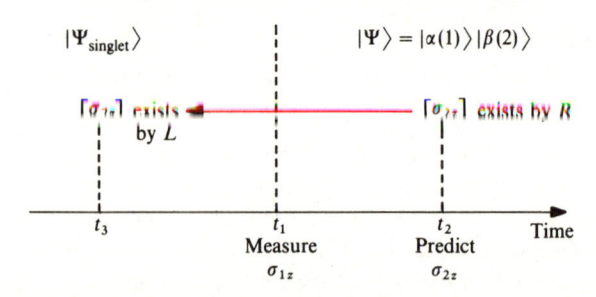

Fig. 8. Schematic illustration of the EPR argument. σ_{1z} is measured at time t_1, and this is used to predict σ_{2z} at a later time t_2. The element of reality $[\sigma_{2z}]$ is asserted to exist at t_2 by the Reality Principle R, and is then projected back to a time t_3 earlier than t_1 by the Locality Principle L. The state of the combined system before the measurement at t_1 is $|\Psi_{\text{singlet}}\rangle$ and after the measurement is $|\Psi\rangle = |\alpha(1)\rangle|\beta(2)\rangle$.

(6) can be rewritten as

$$F \rightarrow \sim (L) \vee \text{Incompleteness} \tag{7}$$

where '$\sim$' denotes negation and '$\vee$' disjunction.

We shall refer to (7) as the *Einstein Dilemma*. It says that, if we accept the formalism F of QM as correct at the purely observational/instrumental level, then either we must give up L the locality principle or we must admit the incompleteness of F. If, with Einstein, we are not prepared to give up L, then (7) constitutes an argument for the incompleteness of F. This indeed is the conclusion of the EPR paper. On the other hand, if we assume that F is complete, then (7) constitutes an argument for nonlocality in the sense that, at the very time t_1 at which the measurement of σ_{1z} took place, an element of reality $[\sigma_{2z}]$ was brought into existence. Since the events comprising the measurement of σ_{1z} and the bringing into existence of $[\sigma_{2z}]$ are simultaneous, relative to the reference frame with respect to which $|\Psi_{\text{singlet}}\rangle$ is the appropriate spin state of the two particles, then these events are at space-like separation, and so L is being violated in the Einstein sense. This way of arguing for nonlocality we shall refer to as the EPR paradox. It is a paradox, not in the strict logical sense, but in the sense of involving a counter-intuitive conclusion, viz. a violation of Einstein locality.

So far we have discussed the EPR argument as it applies to F, the minimal instrumentalist interpretation of the QM formalism. Let us now consider how the argument would look if we first of all filled out the interpretation F with the interpretations we labelled A, B, and C in

The Einstein–Podolsky–Rosen Incompleteness Argument

Chapter 2 (see pp. 45 ff.). These are all intended to provide complete descriptions of reality. So, if by the EPR argument we demonstrate elements of reality *denied to exist by any of these interpretations*, and hence to show its *incompleteness*, we can conclude that the interpretation is actually false.

On view A it is asserted that $[\sigma_{2z}]$ exists at all times. So, there would be no need for the EPR argument to show the existence of $[\sigma_{2z}]$ in the state $|\Psi_{\text{singlet}}\rangle$ and hence the incompleteness of F.

In the case of view B, we can be more specific as to what sort of effects are denied to be possible by the locality principle L. Indeed, let us replace L by the formulation

LOC_1: An unsharp value for an observable cannot be changed into a sharp value by measurements performed 'at a distance'.

Then the EPR argument leads to the result

$$B \wedge \text{LOC}_1 \rightarrow \text{Incompleteness} \qquad (8)$$

$$\rightarrow \sim(B) \qquad (9)$$

From (9) we conclude

$$B \rightarrow \sim(\text{LOC}_1) \qquad (10)$$

i.e. view B in conjunction with the EPR argument leads to a demonstration that LOC_1 is violated in the Einstein sense. Alternatively, we can read (10) contrapositively as showing that if we assume LOC_1 then view B is refuted.

Let us now turn to view C. Here the appropriate specialization of L is

LOC_2: A previously undefined value for an observable cannot be defined by measurements performed 'at a distance'.

We then obtain

$$C \wedge \text{LOC}_2 \rightarrow \text{Incompleteness} \qquad (11)$$

$$\rightarrow \sim(C) \qquad (12)$$

Hence

$$C \rightarrow \sim(\text{LOC}_2) \qquad (13)$$

Now Bohr's response to the EPR argument, and his rejection of the charge of incompleteness, amounts essentially to the claim that LOC_2 can be violated without any physical effects being transmitted in violation of SR. This denial of LOC_2 prevents the use of (13) to argue contrapositively for the falsity of C, and also means from (11) that the argument for incompleteness breaks down, because one of the assumptions, LOC_2, of the argument is not subscribed to. Clearly this manoeuvre to block the EPR argument depends on our accepting the definability criteria associated with Bohr's complementarity philosophy which has already been discussed in Section 2.3.

We now want to comment briefly on the differences between the argument presented above and the original EPR paper:

1. The Reality Criterion R is given the following formulation, which we call R'.

R': If, without in any way disturbing a system, we can predict with certainty (i.e. with probability equal to unity) the value of a physical quantity, then there exists an element of physical reality corresponding to the physical quantity.

R' is not explicit about the time at which the existence of the element of reality can be inferred, and furthermore it mixes up the locality or nondisturbance principle with the reality criterion, which somewhat obscures the logical structure of the argument for incompleteness.

2. EPR demonstrate the simultaneous existence of elements of reality corresponding to non-commuting observables. Thus, continuing with our spin version of the argument, having demonstrated the existence of $[\sigma_{2z}]$ in the state $|\Psi_{\text{singlet}}\rangle$, we could exploit the fact that $|\Psi_{\text{singlet}}\rangle$ can also be expanded in terms of eigenkets of σ_{1x} and σ_{2x}, as in Eq. (1.103), and the mirror-image correlations of the X-components of the Pauli spin observables for the two particles now enable us to infer the existence of $[\sigma_{2x}]$ in the state $|\Psi_{\text{singlet}}\rangle$. So the two non-commuting observables σ_{2z} and σ_{2x} have elements of reality associated with them in the same QM state. This is perfectly correct, and certainly allows us to infer the incompleteness of the minimal instrumentalist interpretation F of QM. In discussion of the EPR argument, a great deal is often made of the choice presented to the experimenter of measuring either σ_{1z} or σ_{1x}—that, while it is true that both $[\sigma_{2z}]$ and $[\sigma_{2x}]$ are predictable in accordance with the mirror-image correlations built into $|\Psi_{\text{singlet}}\rangle$, they cannot both be predicted, since σ_{1z} and σ_{1x}, being incompatible observables, cannot be measured in the same experiment. However, it should be clear that the argument for incompleteness goes through without any consideration of the alternative possibilities of measuring σ_{1z} or σ_{1x}. As we have shown, we need only consider the single component, σ_{1z} say, to establish incompleteness.

3. In their original paper, EPR do not use the spin version of the argument which is due to Bohm (see notes and references at the end of

The Einstein–Podolsky–Rosen Incompleteness Argument

this chapter). Instead, they consider measurements of position and momentum observables for two particles in one-dimensional motion. Thus they consider the state

$$|\Psi\rangle = \int_{-\infty}^{\infty} |p\rangle|-p\rangle e^{-ilp}\, dp \qquad (14)$$

where we use the convention that, in the tensor product, the first component refers to particle 1 and the second to particle 2. (14) is thus a superposition of simultaneous eigenkets of the momenta P_1 and P_2 of the two particles, with associated eigenvalues p and $-p$ respectively. Hence $|\Psi\rangle$ is itself an eigenket of $P_1 + P_2$ with the eigenvalue zero. But (14) is also an eigenket of $Q_1 - Q_2$, where Q_1 and Q_2 are the positions of the two particles.

This follows at once by expanding

$$|p\rangle = \int_{-\infty}^{\infty} \langle q|p\rangle |q\rangle\, dq$$

$$|-p\rangle = \int_{-\infty}^{\infty} \langle q'|-p\rangle |q'\rangle\, dq'$$

and remembering $\qquad \langle q|p\rangle = \dfrac{1}{\sqrt{2\pi}} e^{iqp}$

and $\qquad \langle q'|-p\rangle = \dfrac{1}{\sqrt{2\pi}} \cdot e^{-iq'p}$

So we obtain

$$|\Psi\rangle = \frac{1}{2\pi} \int_{-\infty}^{\infty}\int_{-\infty}^{\infty}\int_{-\infty}^{\infty} dq\, dq'\, dp\, e^{i(q-q'-l)p}|q\rangle|q'\rangle$$

$$= \int_{-\infty}^{\infty} dq \int_{-\infty}^{\infty} dq'\, \delta(q-q'-l)|q\rangle|q'\rangle$$

$$= \int_{-\infty}^{\infty} |q\rangle|q-l\rangle\, dq \qquad (15)$$

From (15)

$$(Q_1 - Q_2)|\Psi\rangle = l|\Psi\rangle$$

showing that $|\Psi\rangle$ is indeed an eigenket of $Q_1 - Q_2$ with the eigenvalue l.

Clearly, the state $|\Psi\rangle$ as specified in (14) or (15) has built into it similar correlation features to the spin example. Thus (14) shows that measurement of P_1 enables us to predict the result of measuring P_2, viz. the negative of the measurement result for P_1, and similarly (15) shows that measuring Q_1 enables us to predict that the result of measuring Q_2 will be the same value displaced through the fixed distance l. EPR are thus able to conclude that, in the state $|\Psi\rangle$, $[P_2]$ and $[Q_2]$ simultaneously exist, and hence to infer the incompleteness of F.

It is instructive to consider how the state $|\Psi\rangle$ might be prepared. Consider a thin rigid diaphragm with two parallel slits which are very narrow compared with their separation, and through each of which one particle, defined by a wave-packet of dimensions large compared with the width of the slits, passes independently of the other. We suppose that the initial momentum of each particle is sharply defined about some average value which specifies the motion of the incoming particles as being directed perpendicular to the diaphragm. Let the diaphragm be suspended by weak springs from a solid yoke which is rigidly connected to the rest of the spatial frame. In these circumstances we are in a position to 'survey' the transverse momentum communicated between the two particles and the diaphragm, providing that we renounce any control of the precise location of the diaphragm relative to the spatial reference frame. If we now select pairs of particles emerging from the two slits with zero transfer of momentum to the diaphragm, and if Q_1, P_1 and Q_2, P_2 refer to the transverse locations and momenta of the two particles, then the state vector for describing such pairs of particles in respect of their transverse motions is effectively just the EPR state $|\Psi\rangle$ we have been discussing.

Two points should be noticed. First, the tight correlation between measurements of Q_1 and Q_2 will only be true at the instant at which the particles emerge from the slits. The time evolution of $|\Psi\rangle$ will result in a progressively increasing decorrelation between measurement results for Q_1 and Q_2, due to the usual diffraction effects behind a narrow slit. Secondly, $|\Psi\rangle$ is a state for which $Q_1 - Q_2$ has the sharp value l, but for which Q_1 and Q_2 separately have unsharp values, ranging in the limiting situation we are considering from $-\infty$ to $+\infty$. From the point of view of Schrodinger time evolution, the EPR state acts as an infinite line source, which is 'incoherent' in the sense that no interference effects arise between the Schrödinger waves originating at different points of the source.

Notes and References

The original EPR argument was given in Einstein, Podolsky, and Rosen (1935). Bohr's response is to be found in Bohr (1935). Bohm (1951) presents the spin version of the argument. The locality principles LOC$_1$ and LOC$_2$ were introduced in Redhead (1983). The thought experiment for producing the EPR state is given in Bohr (1935). For further discussion see Redhead (1981).

4

Nonlocality and the Bell Inequality

4.1. The Bell Inequality

In the preceding chapter we presented the Einstein dilemma, that the minimal instrumentalist interpretation F of QM either implied nonlocality or that F was incomplete. Einstein, as we have seen, chose the incompleteness horn of this dilemma and concluded that, at any rate in certain states, observables for which these states were not eigenstates nevertheless possessed sharp values. This suggested the programme of 'completing' QM in the style advocated in what we referred to as view A in Chapter 2. View A, it will be recalled, says that all observables, in all states, have sharp values. But then, in a famous paper published in 1964, Bell showed that view A, in conjunction with a locality principle appropriate to view A, which we shall term LOC_3, implied a certain inequality between measurable correlation coefficients in a slight extension of the Bohm spin example for the EPR argument. And this inequality, now usually referred to as the Bell inequality, turns out to be in disagreement, over a certain range of conditions, with the predictions of F itself.

This raises two questions. First there is a logical point, that filling out the interpretation F to the complete interpretation of view A is not consistently possible unless the locality principle LOC_3 used in deriving the Bell inequality is violated. In other words, following the incompleteness horn of the Einstein dilemma has not allowed us to escape nonlocality, but has itself landed us in a violation of LOC_3. The second question is whether the predictions of F for the particular set-up envisaged by Bell actually agree with experiment. It might be that the Bell inequality is not violated by experiment, so that view A and LOC_3 could be maintained, but F itself is what is wrong.

We shall now proceed to develop this circle of ideas in more detail. First, let us state the locality principle, LOC_3, appropriate to view A.

LOC_3:
A sharp value for an observable cannot be changed into another sharp value by altering the setting of a remote piece of apparatus.

Let us now see how view A supplemented by LOC_3 leads to the Bell inequality. In the original proof given by Bell, the hidden-variable version of view A, described in Section 2.1 above, was employed. But, as noted there, this involves additional assumptions over and above the existence of sharp values for all observables in all states. In particular, the hidden-variable approach commits us to the existence of joint probabilities for incompatible observables. This 'hidden' assumption might then be incriminated as responsible for the Bell inequality, leaving the locality assumption unchallenged. It is important, therefore, that a proof of the Bell inequality can be given which does not make use of the hidden-variable machinery, and which makes no assumption of joint probability distributions for incompatible observables. This we proceed to do.

Consider again Bohm's version of the EPR argument. Two spin-$\frac{1}{2}$ particles emerge from a source S in a singlet spin state, and move in opposite directions towards two spin-meters which can measure the spin-projection of either particle along any specified direction. We shall consider two directions or 'settings' for each spin-meter, viz. **a** and **a'** for A and **b** and **b'** for B. For the nth pair of particles emitted from the source, denote by a_n the spin-component of the A-particle (i.e. the particle travelling towards spin-meter A) projected in the direction **a** in units of $\hbar/2$ when the A-meter is set parallel to **a**. Similarly for a'_n, b_n, and b'_n in an obvious notation. QM of course dictates that, in so far as measurements by the spin-meters merely reveal these values, they are restricted always to be ± 1.

The situation is sketched in Fig. 9, where the two settings of each spin-meter are indicated by the possible positions labelled **a**, **a'**, **b**, and

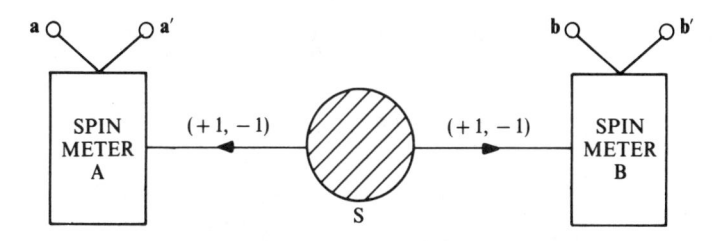

Fig. 9. Schematic illustration of the Bell experiment. S is the source emitting two spin-$\frac{1}{2}$ particles with possible spin-components in any direction of value ± 1 in units of $\hbar/2$. A and B are two spin-meters which can be adjusted to measure spin-components parallel to **a** or **a'** for the A-meter and **b** or **b'** for the B-meter.

b′ of a 'knob' or joystick attached to the corresponding meter. Now form the expression

$$\gamma_n = a_n b_n + a_n b'_n + a'_n b_n - a'_n b'_n \qquad (1)$$

γ_n clearly has integral values which can at most lie between -4 and $+4$ inclusive. The trick here is that the value of the fourth term (with the minus sign) is the product of the first three. Thus

$$a_n b_n \cdot a_n b'_n \cdot a'_n b_n = a_n^2 \cdot b_n^2 \cdot a'_n b'_n = a'_n b'_n$$

A little thought will show that this fact restricts the value of γ_n to ± 2.

This is confirmed by the following simple argument. Write

$$\gamma_n = a_n(b_n + b'_n) + a'_n(b_n - b'_n) \qquad (2)$$

Now b_n and b'_n must have either the same sign or opposite sign. In either case, only one term in (2) is non-vanishing, and its value is clearly ± 2.

Now consider N events and form

$$\left| \frac{1}{N} \sum_{n=1}^{N} \gamma_n \right| = \left| \frac{1}{N} \sum_{n=1}^{N} a_n b_n + \frac{1}{N} \sum_{n=1}^{N} a_n b'_n + \frac{1}{N} \sum_{n=1}^{N} a'_n b_n - \frac{1}{N} \sum_{n=1}^{N} a'_n b'_n \right|$$

But, since $\gamma_n = \pm 2$ for all n, this expression must be less than or equal to 2.

Define correlation coefficients

$$\left. \begin{aligned} c(\mathbf{a}, \mathbf{b}) &= \lim_{N \to \infty} \frac{1}{N} \sum_{n=1}^{N} a_n b_n \\ c(\mathbf{a}, \mathbf{b}') &= \lim_{N \to \infty} \frac{1}{N} \sum_{n=1}^{N} a_n b'_n \\ c(\mathbf{a}', \mathbf{b}) &= \lim_{N \to \infty} \frac{1}{N} \sum_{n=1}^{N} a'_n b_n \\ c(\mathbf{a}', \mathbf{b}') &= \lim_{N \to \infty} \frac{1}{N} \sum_{n=1}^{N} a'_n b'_n \end{aligned} \right\} \qquad (3)$$

Then in the limit as $N \to \infty$ we can conclude

$$|c(\mathbf{a}, \mathbf{b}) + c(\mathbf{a}, \mathbf{b}') + c(\mathbf{a}', \mathbf{b}) - c(\mathbf{a}', \mathbf{b}')| \leq 2 \qquad (4)$$

The inequality (4) is one form of the so-called Bell inequality.

If we remember that the mean values $\overline{a_n}, \overline{b_n}, \overline{a'_n}$ and $\overline{b'_n}$ are all zero for $|\Psi_{\text{singlet}}\rangle$, and the variances $\overline{(a_n - \bar{a}_n)^2}$, etc. are all unity, then the definition of the correlation coefficients given in (3) agrees with the

Nonlocality and the Bell Inequality

usual definition given in statistics, that the correlation coefficient between two statistical variables a_n, b_n is given in general by

$$\frac{\overline{(a_n - \bar{a}_n)(b_n - \bar{b}_n)}}{(\overline{(a_n - \bar{a}_n)^2} \cdot \overline{(b_n - \bar{b}_n)^2})^{\frac{1}{2}}}$$

where we use the bar to denote average or expectation values.

It is now easy to show that, for suitable choice of the direction **a**, **a'**, **b**, and **b'**, the QM predictions for the correlation coefficients violate the Bell inequality. We have in fact already done the necessary calculations in Chapter 1. Thus from Eq. (1.105) we have immediately

$$c(\mathbf{a}, \mathbf{b}) = -\cos\theta_{ab} \qquad (5)$$

where θ_{ab} is the angle between the directions **a** and **b**. Similarly

$$c(\mathbf{a}, \mathbf{b}') = -\cos\theta_{ab'}$$
$$c(\mathbf{a}', \mathbf{b}) = -\cos\theta_{a'b} \qquad (6)$$
$$c(\mathbf{a}', \mathbf{b}') = -\cos\theta_{a'b'}$$

Choose the directions **a**, **a'**, **b**, **b'** to be coplanar and take **a** parallel to **b** and $\theta_{ab'} = \theta_{a'b} = \phi$, say, so $\theta_{a'b'} = 2\phi$ as illustrated in Fig. 10. For this special choice of directions, the Bell inequality will be satisfied by the QM predictions provided

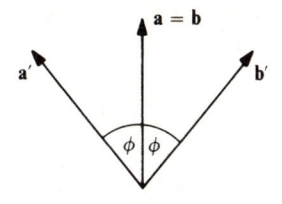

$$F(\phi) \underset{\text{Df}}{=} |1 + 2\cos\phi - \cos 2\phi| \leqslant 2 \qquad (7)$$

Fig. 10. Special choice of directions for illustrating the violation of the Bell inequality.

In Fig. 11 we show $F(\phi)$ plotted as a function of ϕ in the range $0° \leqslant \phi \leqslant 180°$. The Bell inequality is violated for all values of ϕ between $0°$ and $90°$. It is easily checked that the maximum value for $F(\phi)$ is $2\frac{1}{2}$ and is attained for $\phi = 60°$.

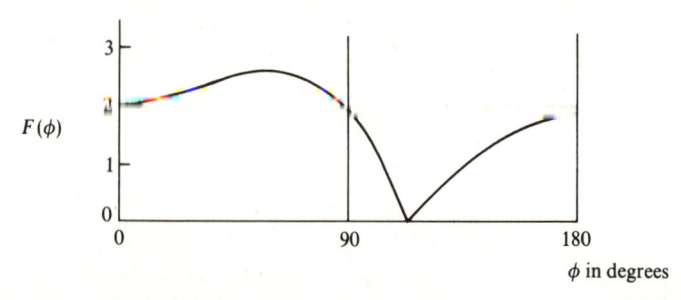

Fig. 11. Graph of $F(\phi)$ given in Eq. (7) against the angle ϕ in the range 0 to 180°. The Bell limit is given by $F = 2$. The inequality is violated for all values of ϕ between 0° and 90°.

It is instructive to consider an example of correlation in classical physics for which the Bell inequality is of course satisfied. Consider two wheels which spin with angular momentum **J** and $-$**J** about their common axle, so that the total angular momentum of the system is zero, just as in the QM spin example we have been discussing. Now let the two wheels fly apart, and measure the sign of the component of each wheel's angular momentum along arbitrary directions **a** for the first wheel and **b** for the second. Now consider an ensemble of such wheel and axle systems with the axles distributed isotropically in space, and let a_n and b_n be the signs of the angular momentum components for the nth axle. So, just as in the QM case, a_n and b_n are always ± 1.

If we draw great circles on the unit sphere whose planes are perpendicular to **a** and **b**, the surface of the sphere is divided into four lunes of aperture θ_{ab}, $\pi - \theta_{ab}$, θ_{ab}, and $\pi - \theta_{ab}$, as illustrated in Fig. 12.

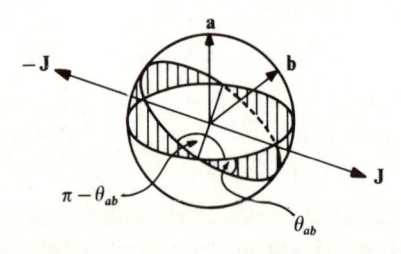

Fig. 12. Classical example of the Bell inequality for two wheels which fly apart with angular momenta **J** and $-$**J**.

If the axle cuts the sphere in the region of the two shaded lunes of aperture θ_{ab}, as shown in the figure, then clearly $a_n b_n$ will be $+1$, while if it cuts the sphere in the region of the two unshaded lunes of aperture $\pi - \theta_{ab}$, $a_n b_n$ will be -1. With an isotropic distribution of axle directions we have then the simple result

$$c(\mathbf{a}, \mathbf{b}) = \overline{a_n b_n} = \frac{2\theta_{ab}(+1) + 2(\pi - \theta_{ab})(-1)}{2\pi}$$

$$= -1 + \frac{2\theta_{ab}}{\pi} \qquad (8)$$

With the choice of the four directions $\mathbf{a}$, $\mathbf{a}'$, $\mathbf{b}$, and $\mathbf{b}'$ shown in Fig. 10, it is easily checked that the LHS of the Bell inequality comes out equal to 2; so in this particular example the Bell inequality is saturated but not, of course, violated.

The reason why the Bell inequality is violated in QM is due to the angular dependence of the correlation coefficients specified in Eqs. (5) and (6). Comparing (5) and (8), notice how for small θ_{ab} the QM prediction 'hangs on' to perfect anti-correlation more tightly than in the classical example. Thus from (5)

$$c(\mathbf{a}, \mathbf{b}) = -\cos\theta_{ab} \simeq -1 + \tfrac{1}{2}\theta_{ab}^2 \ldots$$

So de-correlation is proportional to θ_{ab}^2 rather than θ_{ab}, as specified in (8).

The essential ingredient that has gone into the proof of Bell's inequality is the assumption of LOC_3. For example, we have assumed that the value of a_n is the same whether we are measuring b_n or b'_n, that the change in setting of the knob on the spin-meter B from $\mathbf{b}$ to $\mathbf{b}'$ does not affect the value of a_n, which is 'discovered' by the spin-meter A with knob set in the direction $\mathbf{a}$. This means that both occurrences of a_n in the expression (1) for γ_n have the same value, similarly for a'_n, b_n, and b'_n. This is crucial to the proof that $\gamma_n = \pm 2$. Notice that the definition of a_n does allow for a dependence of the spin-projection of the A-particle parallel to $\mathbf{a}$, on the setting of the spin-meter A.

There are four separate correlation experiments involved in testing the Bell inequality in the form in which we have presented it. These involve combining setting $\mathbf{a}$ with $\mathbf{b}$, $\mathbf{a}$ with $\mathbf{b}'$, $\mathbf{a}'$ with $\mathbf{b}$, and $\mathbf{a}'$ with $\mathbf{b}'$ respectively. We are regarding the four experiments as mutually exclusive, in the sense that each knob can have only one setting for any given experiment; so we are taking here the strong line that incompatible observables cannot be *measured* simultaneously. Nevertheless, we are assuming that a_n, b_n, a'_n, b'_n all have definite *values* which can be measured simultaneously in pairs: a_n with b_n, a_n with b'_n, a'_n with b_n, and a'_n with b'_n.

We illustrate what is going on by considering the table of values for a_n, a'_n, b_n and b'_n (see Fig. 13), in which the four correlation experiments are distinguished as I, II, III, and IV. Each of the four values occurs twice in each row of the table, since it figures in two correlation experiments. The fact that each of the two occurrences has the same numerical value (indicated by the 'ties' in Fig. 13) we term the Matching Condition, which essentially incorporates the assumption of LOC$_3$. Each pair of columns for a given correlation experiment enables one to compute, in the limit as $n \to \infty$, a correlation coefficient with respect to possessed values using all values of n.

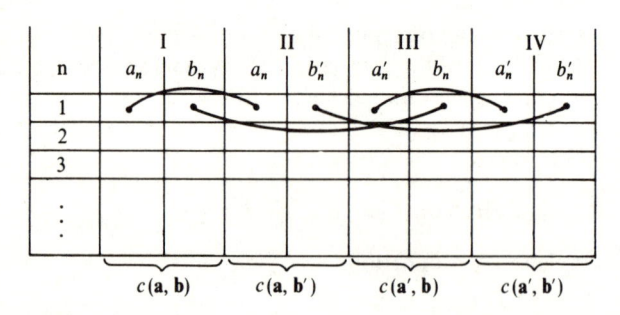

Fig. 13. Schematic table of values for the four correlation experiments I, II, III, and IV in the Bell experiment. The Matching Condition is illustrated by the 'ties' connecting values of the same spin-component in different experiments.

A very important point to notice here is that the possessed values are in general counterfactually possessed. Thus, if the spin-meter A is set parallel to **a** and the spin-meter B parallel to **b**, then the entries for a_n and b_n in the first two columns are actual possessed values. But what about the value for a_n in the third column? This is the value a_n would possess if spin-meter B were set parallel to **b**′ instead of **b**. We shall argue later that these counterfactuals cause no difficulty under the assumption of determinism, i.e. that the values for $a_n, a'_n, b_n,$ and b'_n are deterministically related to the total 'hidden' state of the two particles emerging from the source. But, equally, we shall argue that, if we give up determinism, then the Matching Condition is not licensed

Nonlocality and the Bell Inequality

by appeal to LOC_3. So we have uncovered one 'hidden' assumption in the proof of Bell's inequality, viz. determinism.

But there are other assumptions we need to be explicit about. In order to carry out the correlation experiments we must perform a place selection on the sequence of (in general counterfactually) possessed values which tells us which values of n are to be the subject of which measurement procedure (I, II, III, or IV). We can now isolate two assumptions we have tacitly made:

1. Limiting frequencies computed under these place selections have the same value as those computed with all values of n.
2. The correlation coefficients evaluated with respect to measured values are the same as those evaluated with respect to these selected possessed values.

The first assumption, which we shall term the randomness assumption, is simply that each sequence of a_n's, b_n's, etc. is a random sequence in the Church–von Mises sense, if we suppose that the selection of measurement procedures is governed by some effectively computable rule, and furthermore remains random when we conditionalize on any specified value of properties possessed by the particle entering the opposite wing of the apparatus. This is needed to ensure that sequences such as $\{a_n b_n\}$ are random in addition to $\{a_n\}$ and $\{b_n\}$. The second assumption is justified by a

Principle of Faithful Measurement (FM):
The result of measurement is numerically equal to the value possessed by an observable immediately prior to measurement.

Note that it is conceivable that FM is true, so that every time we make a measurement we reveal what is there, and yet the frequency distribution of measured values might not be equal to the unselected frequency distribution of the possessed values, because measurement selection might skew the underlying distribution. But this would suggest a remarkable conspiracy on the part of nature, that is clearly inconsistent with the experimenter's freedom to choose which possessed values to subject to measurement and, in particular, to specify a rule for selecting measurement procedures.

We can now collect up our various assumptions and write schematically

$$\text{View } A \wedge LOC_3 \wedge \text{Determinism} \wedge$$
$$\text{Randomness} \wedge FM \rightarrow \text{Bell Inequality} \tag{9}$$

Looking at this rag-bag of assumptions made in the proof of the Bell inequality, there may be thought to be a tension or even a downright inconsistency between the assertions of Determinism and Randomness. Determinism said that the values of the spin-projections just before measurement were related deterministically to the state of the two particles produced in the source, while Randomness said that the sequence of successive values of a_n, for example, was a random sequence. But even if states of the source were produced deterministically from previous states of the system, this does not mean that the outcomes of measurement cannot be a random sequence—just that the determination is not ultimately specifiable by a rule which is effectively computable. Randomness in the Church–von Mises sense is compatible with an *ontological* determinism—a given state at one time issues in a *unique* state at a later time—but not with what we may term *pragmatic determinism*, that the prediction of future states can be effectively computed.

We have stressed so far the assumptions that *are* made in the proof of the Bell inequality. But, equally importantly, we stress the assumptions that have *not* been made:

1. We do not assume that the a_n's and a_n''s, for example, have a well-defined joint probability distribution, the correlation functions actually used in deriving the Bell inequality always referring to compatible (commuting) observables. In particular we do not assume that $\frac{1}{N}\sum_{n=1}^{N} a_n a_n'$ has any well-defined limit as $N \to \infty$.
2. We do not assume that a_n and a_n' can be measured simultaneously, contrary to the view expressed by Brody and de la Peña-Auerbach (1979).

4.2. Counterfactuals and Indeterminism

In this section we shall discuss the question of whether we can block the proof of the Bell inequality by giving up determinism.

We shall begin by sketching an argument, due to Stapp and Eberhard, to the effect that a proof can be given of the Bell inequality that is formulated entirely in terms of actual or possible *measurement results*, i.e. the responses of macroscopic measuring apparatus, and which is neutral with regard to possible interpretations of QM such as the views we have labelled A, B, and C. The locality assumption used in this type of argument we shall term LOC_4, which is formulated as follows:

LOC_4:

A macroscopic object cannot have its classical state changed by altering the setting of a remote piece of apparatus.

So the purported theorem is in essence
$$LOC_4 \to \text{Bell Inequality} \qquad (10)$$
If this result could be established, it would provide a very powerful and comprehensive proof of nonlocality in QM. Schematically
$$F \to \sim (\text{Bell}) \to \sim (LOC_4) \qquad (11)$$
and this result would apply just as well to view *B*, for example, as to view *A*. We have already seen (3.10 above) that view *B* implies a violation of LOC_1. But (10) would demonstrate that view *B* also involves a violation of LOC_4. In other words, we would now not need to go through the EPR argument plus the Bell inequality argument to demonstrate nonlocality in QM. Furthermore, the escape from the Einstein dilemma provided by view *C* would also be blocked, since not only LOC_2 would be (acceptably) violated, but also (unacceptably) LOC_4.

Let us then attempt, following the ideas of Stapp and Eberhard, to prove (10). We begin by simply taking over the mathematics of the proof of the Bell inequality given in the preceding section, but with appropriately altered definitions of the symbols a_n, a'_n, b_n, and b'_n. We interpret these quantities now, not as possessed values for the microsystems, but as the responses of the macroscopic spin-meters when set to measure these quantities. Thus a_n = response of *A*-meter when set to measure $\sigma(A) \cdot \mathbf{a}$ for the nth pair of particles emitted by the source. Similarly for a'_n, b_n and b'_n. But since the four correlation experiments I, II, III and IV are mutually exclusive, we must proceed counterfactually.

a_n = response *A*-meter *would* show *if* Experiment I or Experiment II were performed.

Now, the essential crux of the derivation of the Bell inequality was the Matching Condition. Applied to a_n, for example, this says that the result recorded by the *A*-meter would be the same whether Experiment I or Experiment II were performed. But Experiments I and II differ only in the setting of a remote piece of apparatus, spin-meter B. Thus the Matching Condition would follow, and hence the Bell inequality could be proved if the following principle could be sustained.

Principle of Local Counterfactual Definiteness (PLCD):

The result of an experiment which *could* be performed on a microscopic system has a definite value which does not depend on the setting of a remote piece of apparatus.

Clearly then, modulo the randomness assumption involved in the proof,

$$\text{PLCD} \rightarrow \text{Bell Inequality} \qquad (12)$$

Suppose now we could show that

$$\text{LOC}_4 \rightarrow \text{PLCD} \qquad (13)$$

Then from (13) and (12) we would at once obtain the result we are trying to prove, viz. (10).

Arguing contrapositively, (13) says that violation of PLCD implies a violation of LOC_4. This is the crucial claim that we want to examine. Let us consider two simple thought experiments.

In the first, a clock is situated at one end of a table. At time t_2, say 9 o'clock, the clock strikes. I stand at the other hand of the table and at time t_1, say 1 sec. before 9 o'clock, I raise my hand. I now ask the question: 'If I had not raised my hand at time t_1 would the clock have struck at time t_2?' Assuming no mysterious connection between my hand and the mechanism of the clock, the intuitively correct answer to this counterfactual query would seem to be 'Yes', in accordance with PLCD. And, moreover, if the clock had not struck at t_2, when I did not raise my hand at t_1, then I think we could have concluded that the macroscopic behaviour of the clock must depend on the remote setting of my hand, i.e. a violation of LOC_4.

Let us compare this conclusion with what happens in a second thought experiment. In this, the clock is replaced by an atom of radium which decays (emits an α-particle) at time t_2. Again at time t_1, just before t_2, I raise my hand. The question I now ask is: 'If I had not raised my hand at time t_1 would the atom still have decayed at time t_2?' It is not so obvious in this example what the answer to the question should be. Suppose the decay of the radium atom is a truly indeterministic process; then, if I imagine running the course of events through again, with my hand not raised this time, the outcome at time t_2 might just as well be that the atom did not decay. Or, to take a

Nonlocality and the Bell Inequality

slightly different example, suppose I placed a bet on number 17 turning up for a truly indeterministic roulette wheel, and in fact number 16 turns up. Should I correctly say: 'If only I had bet on number 16 I would have won my bet'? This is a conundrum about which philosophers have different views, but it is certainly problematic in a way that is not the case for the deterministic clock example.

One popular way of analysing the truth conditions for counterfactuals is in terms of possible worlds. Let us apply this type of analysis to the counterfactual $\phi \mathbin{\square\!\!\rightarrow} \psi$, where ϕ denotes the condition that I do not raise my hand at time t_1, and ψ the state of affairs that the atom decays at time t_2, and $\mathbin{\square\!\!\rightarrow}$ is a convenient symbol for denoting a counterfactual conditional.

Let the world in which I raise my hand at t_1 and the atom decays at t_2 be denoted by W_i. Let possible worlds W_j for variable j be ordered in respect of 'nearness' to W_i. More specifically, we collect all worlds into classes each of which is composed of worlds 'equidistant' from W_i, and then order these classes in respect of 'distance' from W_i. If W_l is nearer to W_i than W_k, I write $W_l \underset{i}{<} W_k$. Then I analyse $\phi \mathbin{\square\!\!\rightarrow} \psi$ as

$$\exists W_k [\exists W_l ((W_l \underset{i}{<} W_k) \wedge W_l(\phi)) \wedge \forall W_j ((W_j \underset{i}{<} W_k) \rightarrow (W_j(\phi) \rightarrow W_j(\psi)))]$$

thus reducing the counterfactual conditional $\mathbin{\square\!\!\rightarrow}$ in terms of the material conditional $\rightarrow$. $W_p(\phi)$ signifies that ϕ is true in W_p. We are assuming then $\sim W_i(\phi)$. In words: there is a world sufficiently close to W_i such that there exist some worlds closer to W_i in which ϕ holds and such that for any such worlds ψ holds.

This somewhat involved analysis can be illustrated as follows. Represent the classes of equidistant worlds as spheres centred on W_i. Then $\phi \mathbin{\square\!\!\rightarrow} \psi$ would be true in the situation shown in Fig. 14, where the conditions ϕ and ψ are identified with the classes of worlds in which they hold. ϕ is shown shaded and ψ is the region inside the dotted line. W_k is some world on the sphere S_k, and W_j a world on the sphere S_j.

Let us take the example given by David Lewis in his book *Counterfactuals*. If a kangaroo did not have a tail it would fall over. The material conditional is not required to hold for worlds sufficiently far from W_i, i.e. worlds in which the kangaroo has crutches!

In our example time enters in an essential way—we are dealing with a tensed counterfactual. It is important here that the specification of

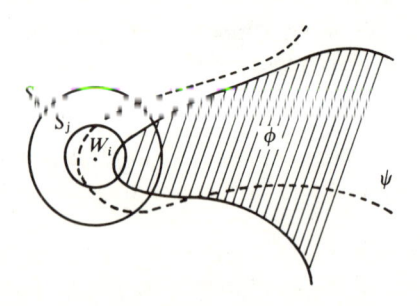

Fig. 14. Truth conditions for the counterfactual conditional $\phi \,\square\!\!\rightarrow\, \psi$. W_i denotes the actual world. S_j and S_k are 'spheres' of possible worlds 'equidistant' from W_i.

'sufficiently close' must refer to states of the world up to but not including t_2—i.e. we must not include the fact of ψ obtaining at t_2 as part of the specification of a sufficiently close world. This proviso is required to deal with the following sort of counter-example: 'If I were to pull the switch at t_1 the hydrogen bomb would explode at t_2.' If I include in the specification of 'sufficiently close world' what happens after the explosion, this statement comes out false, while the statement: 'If I were to pull the switch, the switch would jam' comes out true, since a jammed switch is closer to the actual world than one devastated by the explosion of the hydrogen bomb! This is very counter-intuitive, and can be dealt with as suggested by restricting the specification of 'sufficiently close' to states of the world up to but not including t_2.

Informally, then, we consider a world which differs from our actual world just in the fact that ϕ obtains in the alternative world but not in the actual world; everything else, including the laws of nature, are the same, and we let the world run on to the instant before t_2 and ask: 'Must now ψ occur at t_2?' But if the occurrence of ψ is essentially probabilistic, then there is no necessity for ψ to occur at t_2 in the alternative world. This is not paradoxical. It is just what we mean by saying that the occurrence of ψ is indeterministic (essentially probabilistic). We just cannot refine the description of the world prior to t_2 so as to force the occurrence of ψ at t_2. If such a refinement were possible the occurrence of ψ would be deterministic, not indeterministic.

To return to our original example, the fact that keeping my hand down at t_1 allows the atom not to decay at t_2 has nothing to do with a violation of locality, in particular LOC_4, assuming the decay of the atom is used to trigger some macroscopic recording device. It just involves the recognition of what is meant by the claim that the decay of a radium atom is indeterministic. So what is wrong with PLCD is the meaning attached to 'definite'. The outcome of an essentially indeterministic situation is definite in the sense that on a particular occasion, in a particular world, whatever does happen is the determinate outcome; but it is not definite in the sense that it is both determinate, i.e. necessarily either true or false, and determined, i.e. is rendered either necessarily true or necessarily false, by any possible specification of that world prior to the occurrence of that particular outcome.

The conclusion of this discussion is that, if we assume determinism, PLCD is a valid principle which can indeed be licensed by appeal to LOC_4. Under this assumption, violation of the Bell inequality shows that LOC_4 is violated. Indeed, if one can change the possessed value of some attribute at a distance, and this possessed value is linked deterministically to a macroscopic pointer reading as measurement outcome, then the state of the pointer can be altered 'at a distance'. In other words, LOC_4 holds if and only if LOC_3 holds; so violation of LOC_3 commits us to violation of LOC_4.

But in an indeterministic framework such as that envisaged in view B, for example, it is at least highly questionable whether PLCD can be invoked as a valid principle licensed by an appeal to LOC_4; and if PLCD cannot be used, then the generality claimed by Stapp and Eberhard for their method of proof must be denied.

A final comment on the Stapp–Eberhard proof. There is an alternative way that Eberhard refers to in his (1977) for expressing his results which does not employ counterfactuals. The idea is to record in a table, such as that illustrated in Fig. 13 above, not the results of four correlation experiments which could have been performed, although not simultaneously, but the results of four correlation experiments which are all actually performed. In other words, each pair of columns records a sequence of measurements made with the appropriate pair of knob settings. When four such correlation sequences have been obtained, they are written down side by side to form the complete eight-column table, but each row now refers to four different particle pairs emitted by the source.

So clearly, in general, the Matching Condition will not hold for any particular row. But, by chance, it may. So now perform a place selection on the columns which consists in selecting those rows for which the Matching Condition does hold. If we calculate new correlation coefficients using only these place selected rows of the table, then Eberhard points out that the results cannot in general agree with the correlation coefficients calculated with the whole table. This is perfectly true, since the former, by construction, satisfy the Bell inequality, whereas the latter, in general, violate it. But all this has nothing to do with LOC_4. It just demonstrates the familiar fact that place selections in a random sequence, made in the light of the actual outcomes, can change the limiting frequencies for these outcomes.

There are, however, claims in the literature that sticking with view A but giving up determinism does allow the derivation of the Bell inequality by a *different* line of argument from that used by Stapp and Eberhard. In Section 4.4 we shall consider the case of so-called stochastic hidden-variable theories in relation to questions of nonlocality. But first we shall interrupt the main line of discussion to consider briefly some mathematical manipulations that enable us to give useful alternative formulations of the basic Bell inequality (4).

4.3. Alternative Forms of the Bell Inequality

We start from the inequality (4)
$$|c(\mathbf{a}, \mathbf{b}) + c(\mathbf{a}, \mathbf{b}') + c(\mathbf{a}', \mathbf{b}) - c(\mathbf{a}', \mathbf{b}')| \leqslant 2.$$
By interchanging $\mathbf{b}$ and $\mathbf{b}'$ we have also
$$|c(\mathbf{a}, \mathbf{b}) + c(\mathbf{a}, \mathbf{b}') - (c(\mathbf{a}', \mathbf{b}) - c(\mathbf{a}', \mathbf{b}'))| \leqslant 2.$$
Define
$$X = c(\mathbf{a}, \mathbf{b}) + c(\mathbf{a}, \mathbf{b}') \tag{14}$$
$$Y = c(\mathbf{a}', \mathbf{b}) - c(\mathbf{a}', \mathbf{b}') \tag{15}$$
Then we have shown
$$|X + Y| \leqslant 2 \tag{16}$$
$$|X - Y| \leqslant 2 \tag{17}$$
Suppose X has the same sign as Y. Then using (16)
$$|X| + |Y| = |X + Y| \leqslant 2$$
while if X has the opposite sign to Y, using (17)
$$|X| + |Y| = |X - Y| \leqslant 2$$
So in all cases
$$|X| + |Y| \leqslant 2 \tag{18}$$

Nonlocality and the Bell Inequality

or
$$|Y| \leq 2 - |X| \quad (19)$$

Now consider an experiment for which
$$c(\mathbf{b}, \mathbf{b}) = -1 \quad (20)$$
and choose the direction $\mathbf{a}$ parallel to $\mathbf{b}$ (the notation $c(\mathbf{b}, \mathbf{b})$ just means $c(\mathbf{a}, \mathbf{b})$ when $\mathbf{a}$ is chosen parallel to $\mathbf{b}$). Then (19) becomes

$$|c(\mathbf{a}', \mathbf{b}) - c(\mathbf{a}', \mathbf{b}')| \leq 2 - |c(\mathbf{b}, \mathbf{b}) + c(\mathbf{b}, \mathbf{b}')|$$
$$= 2 - |-1 + c(\mathbf{b}, \mathbf{b}')|$$
$$= 2 - |1 - c(\mathbf{b}, \mathbf{b}')|$$
$$= 2 - (1 - c(\mathbf{b}, \mathbf{b}'))$$
$$= 1 + c(\mathbf{b}, \mathbf{b}')$$

So
$$|c(\mathbf{a}', \mathbf{b}) - c(\mathbf{a}', \mathbf{b}')| \leq 1 + c(\mathbf{b}, \mathbf{b}') \quad (21)$$

for arbitrary directions $\mathbf{b}$, $\mathbf{a}'$ and $\mathbf{b}'$. This is the original form in which Bell presented his inequality. But notice that it is only applicable under the assumption (20).

Although (20) *is* true for the state $|\Psi_{singlet}\rangle$, the inequality (4) is of more general applicability, since it holds in the more general case where the strict anti-correlation expressed in (20) does not apply.

Another useful way of expressing the basic inequality (4) is in terms of probabilities rather than correlation functions. Denote by Prob $(\varepsilon_a, \varepsilon_b)_{a,b}$, for example, the joint probability that measurement of $\sigma(A) \cdot \mathbf{a}$ yields the value ε_a and measurement of $\sigma(B) \cdot \mathbf{b}$ yields the value ε_b, where ε_a and ε_b have the value ± 1 and $\sigma(A)$ and $\sigma(B)$ are as usual the Pauli spin-vectors for the A-particle and the B-particle respectively (i.e. the particles moving towards the A-meter and the B-meter). Similarly for Prob $(\varepsilon_a, \varepsilon_{b'})_{a,b'}$, Prob $(\varepsilon_{a'}, \varepsilon_b)_{a',b}$ and Prob $(\varepsilon_{a'}, \varepsilon_{b'})_{a',b'}$. Also let Prob $(\varepsilon_a)_a$ be the probability of measuring the value ε_a for $\sigma(A) \cdot \mathbf{a}$, irrespective of what is measured on the spin-meter B. Similarly for Prob $(\varepsilon_b)_b$. Then we have

$$\text{Prob}(+1)_a = \text{Prob}(+1, +1)_{a,b} + \text{Prob}(+1, -1)_{a,b} \quad (22)$$
$$\text{Prob}(+1)_b = \text{Prob}(+1, +1)_{a,b} + \text{Prob}(-1, +1)_{a,b} \quad (23)$$

and
$$\text{Prob}(+1, +1)_{a,b} + \text{Prob}(-1, -1)_{a,b} + \text{Prob}(+1, -1)_{a,b}$$
$$+ \text{Prob}(-1, +1)_{a,b} = 1 \quad (24)$$

From Eqs. (22) to (24) we easily derive
$$\text{Prob}(+1, -1)_{a,b} = \text{Prob}(+1)_a - \text{Prob}(+1, +1)_{a,b} \quad (25)$$
$$\text{Prob}(-1, +1)_{a,b} = \text{Prob}(+1)_b - \text{Prob}(+1, +1)_{a,b} \quad (26)$$

and
$$\text{Prob}(-1,-1)_{a,b} = 1 - \text{Prob}(+1,+1)_{a,b} - \text{Prob}(+1,-1)_{a,b}$$
$$- \text{Prob}(-1,+1)_{a,b} = 1 + \text{Prob}(+1,+1)_{a,b} - \text{Prob}(+1)_a$$
$$- \text{Prob}(+1)_b \tag{27}$$

From Eqs. (25) to (27) we obtain
$$c(\mathbf{a},\mathbf{b}) = \text{Prob}(+1,+1)_{a,b} + \text{Prob}(-1,-1)_{a,b}$$
$$- \text{Prob}(+1,-1)_{a,b} - \text{Prob}(-1,+1)_{a,b} = 4\,\text{Prob}(+1,+1)_{a,b}$$
$$- 2\,\text{Prob}(+1)_a - 2\,\text{Prob}(+1)_b + 1 \tag{28}$$

Writing down similar expressions for $c(\mathbf{a},\mathbf{b}')$, $c(\mathbf{a}',\mathbf{b})$ and $c(\mathbf{a}',\mathbf{b}')$, we easily derive the following inequality from (4)
$$-1 \leqslant \text{Prob}(+1,+1)_{a,b} + \text{Prob}(+1,+1)_{a,b'} + \text{Prob}(+1,+1)_{a',b}$$
$$- \text{Prob}(+1,+1)_{a',b'} - \text{Prob}(+1)_a - \text{Prob}(+1)_b \leqslant 0 \tag{29}$$

More generally, if we choose $\varepsilon_a = \varepsilon_{a'} = \varepsilon_A$, say, and $\varepsilon_b = \varepsilon_{b'} = \varepsilon_B$, say, then we can write
$$-1 \leqslant \text{Prob}(\varepsilon_A, \varepsilon_B)_{a,b} + \text{Prob}(\varepsilon_A, \varepsilon_B)_{a,b'}$$
$$+ \text{Prob}(\varepsilon_A, \varepsilon_B)_{a',b} - \text{Prob}(\varepsilon_A, \varepsilon_B)_{a',b'}$$
$$- \text{Prob}(\varepsilon_A)_a - \text{Prob}(\varepsilon_B)_b \leqslant 0 \tag{30}$$

where $\varepsilon_A = \pm 1$, $\varepsilon_B = \pm 1$.

The set of four inequalities comprised in (30) are actually equivalent to (4), in the sense that we can derive (4) from (30) by multiplying the two inequalities for which $\varepsilon_A \neq \varepsilon_B$ by -1 and adding to the two inequalities for which $\varepsilon_A = \varepsilon_B$.

4.4. Stochastic Hidden-Variable Theories

We now revert to the question of whether a proof can be given of the Bell inequality if we combine view A with indeterminism. The framework of so-called stochastic hidden-variable theories will provide a basis for this discussion. (We use the term 'theory' rather than 'interpretation' to allow for the possibility of not reproducing all the empirical predictions of QM.) The idea of such theories is that the 'complete' hidden-variable description of the source does not determine the values of local observables possessed by the two particles in the Bell type of experiment, but only the probabilities for possible values to occur. We can think picturesquely that the values of the spin-components in any given direction are developing in time stochastically, the state of the source controlling only the probabilities that

Nonlocality and the Bell Inequality

particular values will be revealed when subsequent measurements are performed. We will still suppose that faithful measurement is true, i.e. that measurement at time t reveals the value possessed at time t. As we have said before, if measurement results were themselves linked stochastically to possessed values, it would be difficult to know in what sense one could talk of measurement at all. More formally, we shall assume the existence of a joint probability density

$$\text{Prob}\,(\varepsilon_a, \varepsilon_b, \lambda)_{a,b}^{\eta_A, \eta_B}$$

for the values ε_a and ε_b to be possessed by the observables a and b, which are shorthand for $\sigma(A) \cdot \mathbf{a}$ and $\sigma(B) \cdot \mathbf{b}$ respectively, and the value λ for the hidden-variable specifying the state of the source. The superscripts η_A and η_B indicate the settings of the two spin-meters A and B respectively.

If $\eta_A = \mathbf{a}$ and $\eta_B = \mathbf{b}$, then this will be the probability of finding the results ε_a and ε_b on measuring a and b, together with the value λ for the hidden variable. This joint probability in terms of measurement results will be denoted simply by

$$\text{Prob}\,(\varepsilon_a, \varepsilon_b, \lambda)_{a,b}$$

But even if $\eta_A \neq \mathbf{a}$, $\eta_B \neq \mathbf{b}$, the joint probability $\text{Prob}\,(\varepsilon_a, \varepsilon_b, \lambda)_{a,b}^{\eta_A, \eta_B}$ is supposed to exist, although its values will not translate immediately in terms of the probabilities for measurement results.

We now write

$$\begin{aligned}
\text{Prob}\,(\varepsilon_a, \varepsilon_b, \lambda)_{a,b}^{\eta_A, \eta_B} &= \text{Prob}\,(\varepsilon_a/\varepsilon_b\ \&\ \lambda)_a^{\eta_A, \eta_B} \\
&\times \text{Prob}\,(\varepsilon_b/\lambda)_b^{\eta_A, \eta_B} \\
&\times \rho^{\eta_A, \eta_B}(\lambda)
\end{aligned} \qquad (31)$$

where $\text{Prob}\,(\varepsilon_a/\varepsilon_b\ \&\ \lambda)_a^{\eta_A, \eta_B}$ is the conditional probability for a to have the value ε_a, given values ε_b for b and λ for the hidden variable (with settings η_A and η_B for the spin-meters), $\text{Prob}\,(\varepsilon_b/\lambda)_b^{\eta_A, \eta_B}$ is the conditional probability for b to possess the value of ε_b given the value λ for the hidden variable, and $\rho^{\eta_A, \eta_B}(\lambda)$ is the probability density for finding the value λ of the hidden variable.

In order to derive the Bell inequality, we begin by making the following *completeness* assumption

$$\text{Prob}\,(\varepsilon_a/\varepsilon_b\ \&\ \lambda)_a^{\eta_A, \eta_B} = \text{Prob}\,(\varepsilon_a/\lambda)_a^{\eta_A, \eta_B} \qquad (32)$$

The significance of (32), first pointed out by Jarrett (1984), is that λ is sufficient to determine *completely* $\text{Prob}\,(\varepsilon_a/\varepsilon_b\ \&\ \lambda)_a^{\eta_A, \eta_B}$. Specification of

ε_b is not required. We shall return in a moment to discuss the significance of violating the completeness condition.

Under the completeness assumption (32), Eq. (31) reduces to

$$\text{Prob}\,(\varepsilon_a, \varepsilon_b, \lambda)_{a,b}^{\eta_A,\eta_B} = \text{Prob}\,(\varepsilon_a/\lambda)_a^{\eta_A,\eta_B}$$
$$\times \text{Prob}\,(\varepsilon_b/\lambda)_b^{\eta_A,\eta_B} \times \rho^{\eta_A,\eta_B}(\lambda) \qquad (33)$$

But in order to derive the Bell inequality, it is necessary to make the following additional locality assumptions:

$\text{Prob}\,(\varepsilon_a/\lambda)_a^{\eta_A,\eta_B}$ is independent of η_B

$\text{Prob}\,(\varepsilon_b/\lambda)_b^{\eta_A,\eta_B}$ is independent of η_A

$\rho^{\eta_A,\eta_B}(\lambda)$ is independent of η_A and η_B.

Introducing these further locality assumptions into (33), we obtain finally

$$\text{Prob}\,(\varepsilon_a, \varepsilon_b, \lambda)_{a,b}^{\eta_A,\eta_B}$$
$$= \text{Prob}\,(\varepsilon_a/\lambda)_a^{\eta_A}$$
$$\times \text{Prob}\,(\varepsilon_b/\lambda)_b^{\eta_B}$$
$$\times \rho(\lambda) \qquad (34)$$

where we have suppressed those indices on which the indicated probabilities do not depend. In particular, with ε_a and ε_b now referring to measurement results,

$$\text{Prob}\,(\varepsilon_a, \varepsilon_b, \lambda)_{a,b} = \text{Prob}\,(\varepsilon_a/\lambda)_a \times \text{Prob}\,(\varepsilon_b/\lambda)_b \times \rho(\lambda) \qquad (35)$$

The representation (35) is often referred to as 'factorizability' in the literature.

From (35)

$$\text{Prob}\,(\varepsilon_a, \varepsilon_b)_{a,b} = \int_\Lambda \text{Prob}\,(\varepsilon_a/\lambda)_a \cdot \text{Prob}\,(\varepsilon_b/\lambda)_b \cdot \rho(\lambda)\,d\lambda \qquad (36)$$

Similarly

$$\text{Prob}\,(\varepsilon_a, \varepsilon_{b'})_{a,b'} = \int_\Lambda \text{Prob}\,(\varepsilon_a/\lambda)_a \cdot \text{Prob}\,(\varepsilon_{b'}/\lambda)_{b'} \cdot \rho(\lambda)\,d\lambda \qquad (37)$$

$$\text{Prob}\,(\varepsilon_{a'}, \varepsilon_b)_{a',b} = \int_\Lambda \text{Prob}\,(\varepsilon_{a'}/\lambda)_{a'} \cdot \text{Prob}\,(\varepsilon_b/\lambda)_b \cdot \rho(\lambda)\,d\lambda \qquad (38)$$

$$\text{Prob}\,(\varepsilon_{a'},\varepsilon_{b'})_{a',b'} = \int_\Lambda \text{Prob}\,(\varepsilon_{a'}/\lambda)_{a'} \cdot \text{Prob}\,(\varepsilon_{b'}/\lambda)_{b'} \cdot \rho(\lambda)\,d\lambda \quad (39)$$

and
$$\text{Prob}\,(\varepsilon_a)_a = \int_\Lambda \text{Prob}\,(\varepsilon_a/\lambda)_a \cdot \rho(\lambda)\,d\lambda \quad (40)$$

$$\text{Prob}\,(\varepsilon_b)_b = \int_\Lambda \text{Prob}\,(\varepsilon_b/\lambda)_b \cdot \rho(\lambda)\,d\lambda \quad (41)$$

With the representations (36), (37), (38), (39), (40), and (41) it is possible to prove the inequality (30). This follows at once from the following inequality that holds for any real numbers x, y, x', y' that lie in the interval $[0, 1]$

$$-1 \leqslant xy + x'y + xy' - x'y' - x - y \leqslant 0 \quad (42)$$

Substituting
$$x = \text{Prob}\,(\varepsilon_a/\lambda)_a$$
$$y = \text{Prob}\,(\varepsilon_b/\lambda)_b$$
$$x' = \text{Prob}\,(\varepsilon_{a'}/\lambda)_{a'}$$
$$y' = \text{Prob}\,(\varepsilon_{b'}/\lambda)_{b'}$$

and assuming
$$\varepsilon_a = \varepsilon_{a'} = \varepsilon_A$$
$$\varepsilon_b = \varepsilon_{b'} = \varepsilon_B$$

yields (30), on integrating over λ and remembering

$$\int_\Lambda \rho(\lambda)\,d\lambda = 1$$

From (30), as we have seen, the basic inequality (4) follows. We have thus succeeded in giving a proof of (4) that avoids counterfactuals.

The violation of the Bell inequality means that we must give up factorizability. But this circumstance can also be derived without making any use of the Bell inequality, for the singlet state correlations, by the following argument. The factorizable stochastic hidden-variable theories cannot accommodate the existence of strict anti-correlation, as in the original EPR set-up with **a** parallel to **b**. In this case we would require $\text{Prob}(+1, +1)_{a,b} = \text{Prob}(-1, -1)_{a,b} = 0$, but substituting in (36) this requires, for example,

$$\int_\Lambda \text{Prob}(+1/\lambda)_a \cdot \text{Prob}(+1/\lambda)_b \cdot \rho(\lambda) \, d\lambda = 0$$

Since all factors in the integrand are non-negative, this implies

$$\text{Prob}(+1/\lambda)_a = 0 \text{ or } \text{Prob}(+1/\lambda)_b = 0$$

Considering, for example, the first alternative implies $\text{Prob}(-1/\lambda)_a = 1$, but repeating the above argument with $\varepsilon_a = \varepsilon_b = -1$, then gives $\text{Prob}(-1/\lambda)_b = 0$, which in turn implies $\text{Prob}(+1/\lambda)_b = 1$. It is clear, then, that in the case of strict anti-correlation all conditional probabilities $\text{Prob}(\varepsilon_a/\lambda)_a$ and $\text{Prob}(\varepsilon_b/\lambda)_b$ are zero or one, and the theory has collapsed into a deterministic one.

It is true, as we shall see in the next section, that the experimental tests of the Bell inequality that have so far been carried out have employed systems that do not involve the strict anti-correlation of our idealized thought experiment. So in these cases we do require the violation of the Bell inequality to demonstrate the failure of factorizability.

We now want to return to the question, what would be the significance of violating the completeness condition (32)? This would mean that probability distributions of properties possessed by the A-particle would depend in an essential way on what property is possessed by the B-particle. Another way of understanding the completeness condition is that it identifies λ, the complete state of the source, as the common (stochastic) cause of a having the value ε_a and b the value ε_b. Thus Eq. (32) tells us that λ *screens off* ε_b from ε_a. If (32) is violated it is often argued that a and b must exhibit a direct stochastic *causal* link on the grounds that the correlations between a and b can only be accounted for on the basis of stochastic links to a common cause or a direct stochastic causal link. But this conclusion can be questioned if proper account is taken of necessary conditions of robustness required for a direct causal link. By robustness of a causal relation we mean the following: A stochastic causal connection between two physical magnitudes a and b pertaining to two separated systems A and B is said to be *robust* if and only if there exists a class of sufficiently small disturbances acting on $B(A)$ such that $b(a)$ screens off $a(b)$ from these disturbances.

Denoting the disturbance acting on B by d, then the first part of this condition can be rendered formally as

$$\exists D(\forall d \in D(\text{Prob}(a = \varepsilon_a/b = \varepsilon_b \& d) = \text{Prob}(a = \varepsilon_a/b = \varepsilon_b))) \quad (43)$$

A similar condition can be written down for disturbance acting on A.

The requirement of robustness as a necessary condition for a causal relation means that suitably small disturbances of either relata do not affect the causal relation. This is essentially the basis of the mark method for identifying causal processes. The processes propagate small disturbances (marks) in a local event-structure in accordance with the causal law at issue.

We can easily translate this robustness condition so as to apply to the singlet-state correlation we are discussing. Consider possible perturbations of the quantum-mechanical state $|\Psi\rangle$ by disturbances acting on the particle B. In order to make the problem tractable we shall restrict the discussion to coupling of particle B to uniform c-number fields of arbitrary strength, which are switched on for some specified interval of time to provide the perturbation. Let Ψ' be a variable ranging over these perturbed states. Then a necessary condition that a and b exhibit a stochastic causal connection for arbitrary choice of the directions $\mathbf{a}$ and $\mathbf{b}$ is:

$$\exists D \forall a \forall b (\forall |\Psi'\rangle \in D \, (\text{Prob}^{|\Psi'\rangle}(a = \varepsilon_a/b = \varepsilon_b)$$
$$= \text{Prob}^{|\Psi\rangle}(a = \varepsilon_a/b = \varepsilon_b))) \quad (44)$$

where the superscript on Prob denotes the quantum-mechanical state, and where the class D is some non-empty set of perturbed states arising from sufficiently weak perturbing fields.

We shall now show by explicit calculation that condition (44) is violated. Since (44) is a necessary condition for a direct stochastic causal link between a and b it will follow that no such link exists.

Denote the spin-projection of σ_A and σ_B along the arbitrarily chosen Z-axis by σ_{Az} and σ_{Bz} respectively. Then we have

$$|\Psi\rangle = \frac{1}{\sqrt{2}}(|\sigma_{Az} = +1\rangle|\sigma_{Bz} = -1\rangle - |\sigma_{Az} = -1\rangle|\sigma_{Bz} = +1\rangle) \quad (45)$$

which is a vector in $\mathbb{H}_A \otimes \mathbb{H}_B$, the tensor product of the Hilbert Spaces, $\mathbb{H}_A$ and $\mathbb{H}_B$ for the particles A and B.

Consider an arbitrary perturbation acting on $\mathbb{H}_B$. It will induce a 2×2 unitary transformation on all vectors belonging to $\mathbb{H}_B$. This transformation is thus an element of $U(2)$, the group of 2-dimensional unitary transformations. It is well-known that $U(2)$ can be exhibited as the direct product of $U(1)$, the group of 1-dimensional unitary transformations, and $SU(2)$, the group of 2-dimensional unitary unimodular transformations. Formally

$$U(2) = U(1) \times SU(2) \qquad (46)$$

Now an element of $U(1)$ merely induces a phase-shift which does not change the physical state (ray) associated with particle B and can be ignored in computing all probabilities. The action of an element of $SU(2)$ can always be represented as $e^{i(\sigma_B \cdot \mathbf{n})\phi/2}$, where the direction of the unit vector $\mathbf{n}$ and the magnitude of the angle ϕ range over the three parameters of the group (note that $0 \leq \phi < 4\pi$).

We denote $e^{i(\sigma_B \cdot \mathbf{n})\phi/2}$ by $u(\mathbf{n}, \phi)$. Then the most general perturbed state is given by

$$|\Psi'\rangle = \frac{1}{\sqrt{2}}(|\sigma_{Az} = +1\rangle u(\mathbf{n}, \phi)|\sigma_{Bz} = -1\rangle - |\sigma_{Az} = -1\rangle u(\mathbf{n}, \phi)|\sigma_{Bz} = +1\rangle) \qquad (47)$$

In what follows we shall consider the particular choice $\varepsilon_a = \varepsilon_b = 1$. Then we have the familiar results (cf. Eq. 1.106)

$$\text{Prob}^{|\Psi\rangle}(a = 1) = \tfrac{1}{2} \qquad (48)$$

$$\text{Prob}^{|\Psi\rangle}(b = 1) = \tfrac{1}{2} \qquad (49)$$

$$\text{Prob}^{|\Psi\rangle}(a = 1/b = 1) = \sin^2 \tfrac{1}{2}\theta_{ab} \qquad (50)$$

where θ_{ab} is the angle between the directions $\mathbf{a}$ and $\mathbf{b}$. We are interested now in calculating $\text{Prob}^{|\Psi'\rangle}(a = 1)$, $\text{Prob}^{|\Psi'\rangle}(b = 1)$ and $\text{Prob}^{|\Psi'\rangle}(a = 1/b = 1)$.

The robustness condition for stochastic casuality is simply

$$\text{Prob}^{|\Psi\rangle}(a = 1/b = 1) = \text{Prob}^{|\Psi'\rangle}(a = 1/b = 1) \qquad (51)$$

We shall show that for any given disturbed state $|\Psi'\rangle$ there always exists directions $\mathbf{a}$ and $\mathbf{b}$ for which (51) is violated.

To calculate the new probabilities apply the unitary transformation $u(\mathbf{n}, -\phi)$ ($= u^{-1}(\mathbf{n}, \phi)$) to the space $\mathbb{H}_B$. This converts $|\Psi'\rangle$ back into $|\Psi\rangle$, but induces a rotation $R(\mathbf{n}, -\phi)$ in the operator σ_B. $R(\mathbf{n}, \phi)$ is an element of the 3-dimensional rotation group $SO(3)$, corresponding to an (active) clockwise rotation about the direction $\mathbf{n}$. The above result is a direct expression of the famous homomorphism that exists between $SU(2)$ and $SO(3)$. ($SU(2)$ is just the simply connected

Nonlocality and the Bell Inequality

universal covering group of $SO(3)$.) Succinctly

$$u(\mathbf{n}, \phi)\sigma_B u(\mathbf{n}, -\phi) = R(\mathbf{n}, \phi)\sigma_B \tag{52}$$

On the left of Eq. 52, u acts on the spinor indices of σ_B, while on the right R acts on the vector indices. Thus the effect of our unitary transformation on the operators a and b is as follows:

$$a = \sigma_A \cdot \mathbf{a} \to a' = a \tag{53}$$

$$b = \sigma_B \cdot \mathbf{b} \to b' = (u(\mathbf{n}, -\phi)\sigma_B u(\mathbf{n}, \phi)) \cdot \mathbf{b}$$
$$= (R(\mathbf{n}, -\phi)\sigma_B) \cdot \mathbf{b}$$
$$= \sigma_B \cdot (R(\mathbf{n}, \phi)\mathbf{b}) = \sigma_B \cdot \mathbf{b}' \tag{54}$$

where

$$\mathbf{b}' = R(\mathbf{n}, \phi)\mathbf{b} \tag{55}$$

Hence we have at once the following results:

1. $$\text{Prob}^{|\Psi'\rangle}(a = 1) = \text{Prob}^{|\Psi\rangle}(a' = 1) = \text{Prob}^{|\Psi\rangle}(a = 1) = \tfrac{1}{2} \tag{56}$$

This shows that the perturbation acting on particle B cannot be used to send signals to the location of particle A. We shall discuss this more fully in Section 4.6 below.

2. $$\text{Prob}^{|\Psi'\rangle}(b = 1) = \text{Prob}^{|\Psi\rangle}(b' = 1) = \tfrac{1}{2} \tag{57}$$

This is a rather surprising result that is a special property of the example under discussion.

3. $$\text{Prob}^{|\Psi'\rangle}(a = 1/b = 1) = \text{Prob}^{|\Psi\rangle}(a' = 1/b' = 1)$$
$$= \text{Prob}^{|\Psi\rangle}(a = 1/b' = 1) = \sin^2 \tfrac{1}{2}\theta_{ab'} \tag{58}$$

The robustness condition (51) thus reduces to $\sin^2 \tfrac{1}{2}\theta_{ab'} = \sin^2 \tfrac{1}{2}\theta_{ab}$, or

$$\theta_{ab} = \theta_{ab'} \tag{59}$$

since the angles all lie in the range 0 to π. (59) can be given a simple geometrical interpretation.
In Fig. 15

ON represents the unit vector $\mathbf{n}$
OB   "   "   "   " $\mathbf{b}$
OB'   "   "   "   " $\mathbf{b}'$

B and B' lie on a circle with centre O' whose plane is perpendicular to ON. $\angle BO'B' = \phi$.

Now $\theta_{ab} = \theta_{ab'}$ is equivalent to
$\cos \theta_{ab} = \cos \theta_{ab'}$
or $\mathbf{a} \cdot \mathbf{b} = \mathbf{a} \cdot \mathbf{b}'$
or $\mathbf{a} \cdot (\mathbf{b} - \mathbf{b}') = 0$

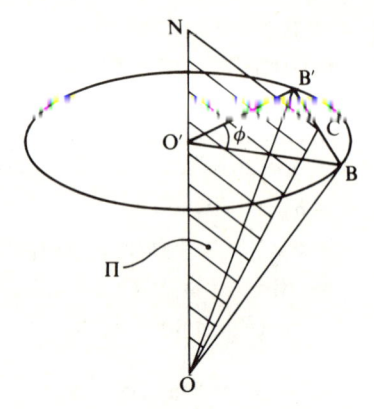

Fig. 15. Rotation of unit vector $\overline{OB}$ through angle ϕ about axis $\overline{ON}$ into new position $\overline{OB'}$. OC is the bisector of the angle between $\overline{OB}$ and $\overline{OB'}$. The plane NOC, denoted by Π, is shown shaded.

Hence **a** must line in the plane perpendicular to the line BB'. Call this plane Π. Π can equally well be characterized as the plane through ON and OC, where OC is the bisector of the angle between **b** and **b'**. The plane Π is shown shaded in the diagram. So we have the following

Theorem:

For any given perturbation on particle B that issues in a rotation of **b** to **b'**, the conditional probability $\text{Prob}^{|\Psi\rangle}(a = 1/b = 1)$ will be invariant (robust) under the perturbation if and only if the direction **a** lies in the plane defined by the axis of rotation and the bisector of the directions **b** and **b'**.

Corollaries:

(1) If **n** coincides with **a**
    $\text{Prob}^{|\Psi\rangle}(a = 1/b = 1)$ is robust for all **b**.
(2) If **n** coincides with **b**
    $\text{Prob}^{|\Psi\rangle}(a = 1/b = 1)$ is robust for all **a**.
(3) For any perturbation (rotation) however small there always exist directions **a** and **b** for which $\text{Prob}^{|\Psi\rangle}(a = 1/b = 1)$ is *not* robust.

It is this last corollary which demonstrates, I believe, that a and b cannot be regarded as related by stochastic causality. The correlations between a and b are a property of the particular quantum-mechanical state, viz. the singlet state, in which the particles emerge from the source. The state involves a feature of holism or nonseparability, which, lacking the necessary robustness for stochastic causality, may

Nonlocality and the Bell Inequality

be termed *passion-at-a-distance* as opposed to *action-at-a-distance*. The A-particle does not possess independent properties (propensities) of its own. The conditional probability Prob $(r_a/r_b)\ \eta_A.\eta_B$ is a candidate for an inherently relational property of the joint two-particle system. The situation should be contrasted with the result of violating the additional locality assumptions introduced below Eq. (33). If these conditions were not satisfied we would have a clear case of action-at-a-distance. Changing the setting of the apparatus B for example would alter the conditional probabilities of properties manifested at the location of the A-particle and the source, and so on.

We shall find a striking analogy to the distinction drawn here between locality in the sense of no-action-at-a-distance and separability in the sense of ascribing properties separately to the two particles, in the developments dealt with in Chapter 6.

4.5. Experimental Tests of the Bell Inequality

In this section we turn to the experimental tests of the Bell inequality that have been carried out during the past fifteen years or so.

The first point we would like to stress is that, since the Bell inequality involves experimentally accessible correlation coefficients, it can be tested directly, without any intermediate reference to QM. As we have seen, the predictions of QM do in certain circumstances violate the Bell inequality, but there are two distinct questions:

1. Is the Bell inequality violated?
2. Does the violation conform to the predictions of QM?

The broad consensus of the experimental results is that both these questions are answered in the affirmative.

There are basically three types of experiment which have, so far, been carried out. The first type measures polarization correlations between two 'visible' photons emitted in a cascade transition from the excited state of an atom such as calcium or mercury, or, in the most recent example, simultaneously from excited deuterium; the second type looks at spin correlations in low-energy proton–proton scattering; while the third type involves γ-ray polarizations from the annihilation decay of the singlet state of positronium. Before discussing the special features of these types of experiments we have listed in Table 1 the principal experiments in each class, and whether the result agrees (indicated by a tick) or disagrees (indicated by a

Table 1. Experimental tests of the Bell inequality. Violation indicated by a tick, no violation by a cross.

Visible Photon Correlation Experiments

Date	Experimenters	Result	Remarks
1972	Freedman and Clauser	✓	Used Ca cascade
1972	Holt and Pipkin	×	Used Hg cascade
1976	Clauser	✓	Used Hg cascade
1976	Fry and Thomson	✓	Used Hg cascade
1981	Aspect, Grangier, and Roger	✓	Used Ca cascade with single-channel polarizers.
1982	Aspect, Grangier, and Roger	✓	Used Ca cascade with 2-channel polarizers.
1982	Aspect, Dalibard, and Roger	✓	Used Ca cascade with optical switches.
1985	Perrie, Duncan, Beyer, and Kleinpoppen.	✓	Used simultaneous 2-photon emission by metastable atomic deuterium

Low Energy Proton–Proton Scattering:

1976	Lamehi-Rachti and Mittig	✓	Used double scattering to measure spin correlations.

γ-Ray Polarization Correlation Experiments

Date	Experimenters	Result	Remarks
1974	Faraci, Gutkowski, Notarrigo, and Pennisi	×	Used Compton polarimeter and ^{22}Na source
1975	Kasday, Ullman, and Wu	✓	Used Compton polarimeter and ^{64}Cu source
1976	Wilson, Lowe, and Butt	✓	Used Compton polarimeter and ^{64}Cu source
1977	Bruno, d'Agostino, and Maroni	✓	Used Compton polarimeter and ^{22}Na source

cross) with the prediction of QM and the violation of the Bell inequality.

We now make some brief comments on these experiments. Only two of them, Holt and Pipkin, and Faraci *et al.*, have disagreed with QM and indeed have shown agreement with the Bell inequality where QM predicts a violation. Of course, from a physicist's point of view, any violation of the predictions of QM is looked at very critically, and in both of the discordant cases the experiments were repeated, and the discrepancies with QM could not be reproduced. On balance, it is now generally agreed that these anomalous experiments must have involved some unsuspected systematic error.

The experiments involving γ-ray polarizations suffer from the necessity of using a secondary Compton scattering as an indicator of the polarization state of the γ-ray. In analysing these experiments, it is assumed that the Klein–Nishina formula is applicable to the Compton scattering even in 'hidden' states of polarization. The same reservations apply in the case of the low energy proton–proton scattering where the scattering of the protons off a secondary target is used as an indicator of their spin orientations.

In the optical photon experiments, one has the big advantage of being able to detect polarizations directly. Typically, pile-of-plates analysers are used to transmit photons of a given linear polarization with high efficiency. With this type of analyser, the orthogonal component of polarization is detected by absence of transmission. The only experiment in which two-channel analysers have been employed is the one by Aspect, Grangier, and Roger (1982). Here a calcite crystal is used to split orthogonal polarizations into two beams, each of which has a detector placed behind it. In all of these optical photon experiments, the photomultipliers used in the detection process are relatively inefficient; and in addition the nature of the angular correlation is such that the experiment is not sensitive to all the coincidence photons (the cascade transition is essentially a three-body decay due to the recoil of the atom). For both these reasons, one is only observing a sample of all the photon pairs emitted from the source; and in assessing the violation of the Bell inequality, it is necessary to assume that the sample of pairs actually observed is not biased in such a way as to explain the violation. For example, since the detection process is 'downstream' from the analysers, it is not surprising that, with suitable *ad hoc* assumptions as to how the efficiency of detection might depend on polarization, one can 'explain' the violation of the Bell inequality in a 'local' way. Such 'explanations' have to be ruled out by auxiliary assumptions concerning the functioning of the apparatus used to analyse and detect the photons. It would be nice to design an experiment which would eliminate the need for such auxiliary assumptions. Such an experiment has been suggested by Lo and Shimony (1981), which employs the coincidence detection of the two dissociation fragments of a metastable molecule. With such a two-body decay, strong angular correlations would obtain, and with Stern–Gerlach analysers and ionization detectors very high efficiencies can in principle be achieved. This experiment appears to be quite possible, and there is no doubt it should be carried out in order to eliminate the 'auxiliary assumptions' loophole in the existing experiments.

There is, however, another and more profound problem raised by all the experiments so far described. They are *static* experiments—that is to say, the choice of analyser setting is made well in advance of the emission of the particles from the source. In principle, there is a possibility that the settings of the analysers 'communicates' itself to the source in such a way as to affect the correlations being measured with these particular settings. This would be an example of violating Bell locality but not Einstein locality, in the terminology introduced in Chapter 3. In order to check for a violation of Einstein locality, one must devise an experiment in which the settings of the analysers are decided after the photons have left the source in the photon-cascade type of experiment. Such an experiment was planned by Aspect and finally carried out by Aspect, Dalibard, and Roger in 1982. In the Aspect experiment, an 'optical' switch or commutator is introduced in the path of each photon which can deflect the photon towards one or other of two analysers with different settings. The arrangement is shown schematically in Fig. 16. The analyser settings are labelled **a**, **a**′, **b** and **b**′ to conform with the arrangement in the idealized Bell experiment shown in Fig. 9. The switch is pulsed at such a frequency that it changes the selection of analyser while the photon is in transit from the source.

Consider the light-cone structure for the source *S* and the two switches *A* and *B*, as seen from the reference frame in which *S*, *A*, and

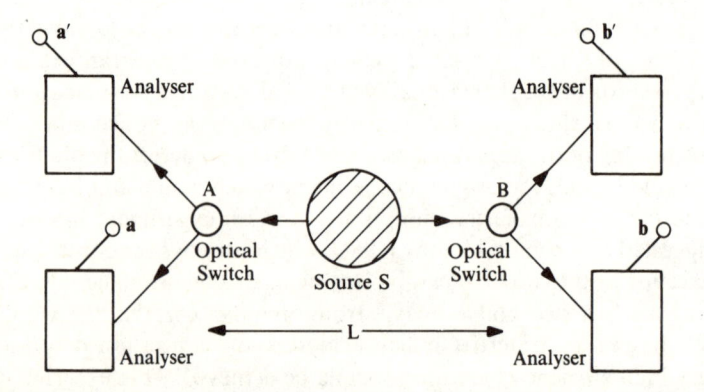

Fig. 16. The Aspect version of the Bell experiment. *A* and *B* are optical switches sending each photon to one or other of a pair of analysers, with settings denoted by **a**, **a**′, **b** and **b**′.

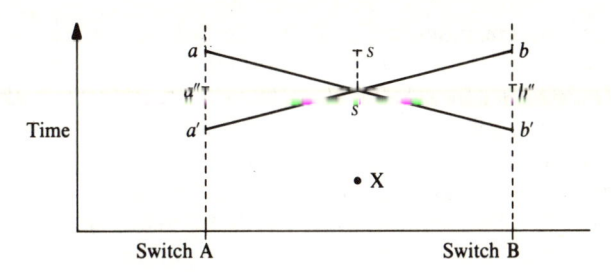

Fig. 17. Light-cone structure for the source S and the two switches A and B in the Aspect experiment. For details see text.

B are all stationary. This is illustrated in Fig. 17. The source emits the two photons at space-time point s'. The photons reach the switches at space-time locations a and b, at which time the source is at s. The part $b'b$ of the world-line of the switch B is outside the light-cone with vertex at s'. Similarly the part $a'a$ of the world-line of switch A is outside the light-cone at s'.

Let us put in some numbers. In Aspect's experiment, the distance L between A and B is 12 metres. Hence it is easy to calculate that $a'a = b'b = 40$ ns (1ns = 10^{-9} seconds). Furthermore, $s's = 20$ ns. The switch is designed to change its setting every 10 ns. This is half the time (20 ns) it takes for the photons to go from the source to the switches. So the switches will certainly alter their settings while the photons are in transit, e.g. at points such as a'' and b'' outside the light-cone at s'. Indeed, in the following discussion we shall assume that a'' and b'' are the last switching operations prior to a and b respectively. Clearly, the photon moving from s' to a is always outside the light-cone with vertex at b'', and similarly the photon moving from s' to b is always outside the light-cone with vertex at a''. In other words, any influence of the switching event at a'' on the polarization state of the photon moving to b, or of the switching event at b'' on the polarization state of the photon moving to a, would be a violation of Einstein locality.

It is worth noting that the emission event involved in the cascade actually has a half-life of about 5 ns, so the location of the source at s' is 'blurred' by this amount. It is important for the above argument that this time is small compared with the 40 ns extent of the exterior region of the light-cone at s', sectioned at the switches A and B.

112 Incompleteness, Nonlocality, and Realism

In the experiment devised by Aspect, the optical switches consist of a glass cell filled with water, in which ultrasonic standing waves are produced via electroacoustic transducers connected to a 25 MHz generator. The cell acts as a variable diffraction grating. When the standing wave is of maximum amplitude, the incoming photon suffers a Bragg reflection from the antinodal planes. Half a period later, when the amplitude of the standing wave is zero, the photon travels straight through the cell without any Bragg reflection (diffraction). The device is illustrated in Fig. 18. In the experiment the Bragg angle θ_B is about $1/4°$, so the deflection $2\theta_B$ of the diffracted beam is approximately $\frac{1}{2}°$. With a 25 MHz generator the switching frequency is 50 MHz, since clearly the switch operates at twice the acoustic frequency. The half-period for switching, i.e. the time interval between the two 'directions' of the switch, comes out at 10 ns, as stated above.

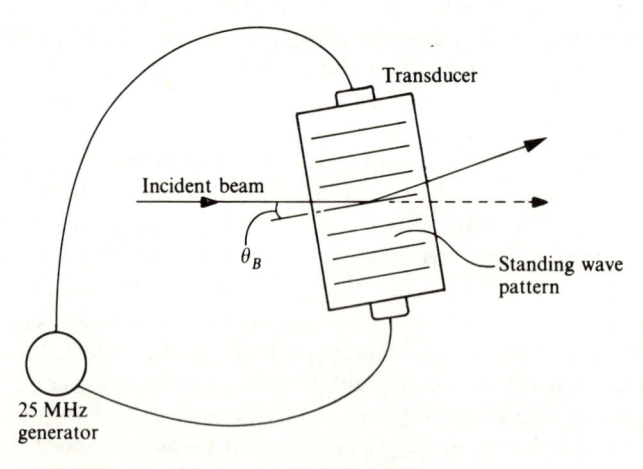

Fig. 18. Optical switch in the Aspect experiment.

The results of Aspect's experiment were a clear violation of the Bell inequality. So prima facie we have an important result here, showing a violation of the Einstein version of LOC_3. There are, however, two points to be noticed:

1. Although the two switches are run from separate independent generators, it is clear that the switches are not truly operated in a random fashion. In other words, referring again to Fig. 17, the state of the apparatus prior to b', i.e. inside the backward light-cone, with vertex at s', determines the switching operation at b''. So knowledge of

Nonlocality and the Bell Inequality

the switch setting at b, when the photon hits the switch, is available at locations inside the backward light-cone at s', and hence could be 'communicated' to s' without violation of Einstein locality. Similarly for the A-switch.

2. But even if attempts were made to randomize the switching operations, this would still be consistent with ontological determinism (compare the discussion on p. 90 above); and some event X in the overlapping backward light-cones of b'' and s' could be held responsible both for the switch-change at b'' and an effect on the source at s', so correlating the state of the switch at b with the state of the source without violation of Einstein locality.

While admitting that bizarre possibilities of this sort could circumvent the demonstration of a violation of Einstein locality in the Aspect experiment, the situation is rather like that referred to in connection with the auxiliary assumptions, which of course also have to be invoked in the Aspect experiment. Duhemian 'good sense' may dictate that we should accept the demonstration in the Aspect experiment of a violation of Einstein locality. The difference lies in the fact that experiments can in principle be designed that do not need the 'auxiliary assumptions', while no experiment can rule out the type of defence of Einstein locality referred to above.

4.6. Statistical Nonlocality

In Section 4.1 we discussed the condition under which a Bell-type experiment might lead us to argue for a violation of LOC_3—that sharp values and hence, assuming FM, measurement results at one location can be changed by altering the setting of a remote piece of apparatus. But LOC_3 is concerned with what happens on a particular occasion. It still leaves open the question of whether the statistical frequencies with which measurement results turn up at one location can be changed by performing different sorts of measurement at another remote location. To deal with this question we introduce still another sense of locality:

LOC_5:

The statistics (relative frequencies) of measurement results of a quantum-mechanical observable cannot be altered by performing measurements at a distance.

Notice that LOC_5 is formulated in terms of the statistics of measurement results, the very question which the formalism of QM, via the statistical algorithm, is designed to answer. If LOC_5 were violated, this would show that the formalism itself, with the minimal instrumentalist interpretation, exhibited nonlocal features, independently of any more comprehensive interpretation of the formalism.

It is therefore important to realize that LOC_5 is not violated in the Bell type of experiment. To see why this is so, let us consider that we decide to measure σ_{1z} on particle 1 in the state $|\Psi_{\text{singlet}}\rangle$ as given in Eq. (3.2). We investigate how this would affect the statistics of any spin-component measurements made on particle 2. We use the ideal measurement theory developed in Section 2.4.

We label the measuring device for σ_{1z} as system 3. Let the initial state of the apparatus be denoted by $|w_0(3)\rangle$ and the final state be denoted by $|w_+(3)\rangle$ if σ_{1z} has the measured value $+1$, and by $|w_-(3)\rangle$ if σ_{1z} has the measured value -1.

Then the state of the whole system 1, 2, and 3 goes from

$$|\Psi_i\rangle = \frac{1}{\sqrt{2}}(|\alpha(1)\rangle|\beta(2)\rangle - |\beta(1)\rangle|\alpha(2)\rangle)|w_0(3)\rangle$$

before the measurement to

$$|\Psi_f\rangle = \frac{1}{\sqrt{2}}(|\alpha(1)\rangle|\beta(2)\rangle|w_+(3)\rangle - |\beta(1)\rangle|\alpha(2)\rangle|w_-(3)\rangle)$$

after measurement.

In terms of von Neumann statistical operators, it follows, from the discussion given in Section 2.4, that $P_{|\Psi_i\rangle}$ behaves like $\frac{1}{2}P_{|\alpha(2)\rangle} + \frac{1}{2}P_{|\beta(2)\rangle}$ in respect of measurements of any observables pertaining to particle 2 only (compare Eq. (2.10)). This arises because of the orthogonality of the states $|\alpha(1)\rangle$ and $|\beta(1)\rangle$. But $P_{|\Psi_f\rangle}$ also behaves like $\frac{1}{2}P_{|\alpha(2)\rangle} + \frac{1}{2}P_{|\beta(2)\rangle}$ in respect of measurements of an observable pertaining to particle 2 only. This is because the interference terms arising in any measurement statistics for particle 2 are now 'killed' twice over by the orthogonality of $|\alpha(1)\rangle$ and $|\beta(1)\rangle$, and also by the orthogonality of $|w_+(3)\rangle$ and $|w_-(3)\rangle$. So the statistics of measurement results pertaining only to particle 2 is unaffected by 'hooking on' the apparatus for measuring σ_{1z} on particle 1. This result holds *a fortiori* if we assume that the measurement actually produces a final *mixed* state for the joint system.

We are referring here of course to the nonselective stage of measurement. If we select a sub-ensemble of particle 2's with

$\sigma_{2z} = +1$, say, this will be described by the statistical operator $P_{|\alpha(2)\rangle}$, which of course gives different statistics in general from $\frac{1}{2} P_{|\alpha(2)\rangle} + \frac{1}{2} P_{|\beta(2)\rangle}$. This selection could be effected in the light of knowing which measurement results we had obtained for σ_{1z} in view of the mirror-image correlations built into $|\Psi_{\text{singlet}}\rangle$. But then the selection is made at the wrong location to provide any 'instantaneous statistical effects' at a distance.

Thus suppose we are measuring a sequence of values for σ_{2z} on successive particles emitted by the source. The sequence might be $+ - - + - + - + + - - - \ldots$ where the limiting frequency of $+$ and $-$ is $\frac{1}{2}$. Suppose we perform measurements of σ_{1z} simultaneously on particle 1. This enables us to 'tag' each particle 2 as either $+$ or $-$ in our abbreviated notation for values of $\sigma_{2z} = \pm 1$. But the tagging information is in the wrong place to change the statistics at the location of particle 2. To do this we would have to transmit the tagging information from location 1 to location 2, with instructions, for example, to insert an absorbing screen every time a minus particle is approaching the spin-meter for particle 2, and to remove it every time a plus particle is approaching. In this way we would clearly change the sequence of observed measurement results of σ_{2z} to $+ + + + + \ldots$, but to effect this change we have to transmit information from location 1 to location 2; we cannot do it simply by 'hooking on' the apparatus to measure σ_{1z}.

A simple example can illustrate the problem. When I lecture in Oxford, the audience there learn instantaneously that my room in London is empty; but to produce a physical change in London, for example to prevent students knocking on my door, the information that I have arrived in Oxford must be transmitted back to London.

The conclusion of this discussion is that the nonlocality putatively demonstrated in the Bell type of experiment cannot be used to transmit information instantaneously between two remote locations. In brief, there is no such thing as a 'Bell' telephone! In this respect the situation is quite different from what would obtain if the operators referring to particles 1 and 2 failed to commute. In such a case LOC_5 would be violated. But the nonlocality demonstrated in the EPR and Bell arguments is subtler than this. In particular, the fact that no statistical effects get transmitted 'at a distance' means that no nonlocality problems arise in an ensemble or statistical interpretation of QM. It is only in the context of an attempt to impute states to individual systems that the difficulties are manifested.

We have discussed the no-signalling result in the context of interactions which perform a measurement on particle 1, and for the particular case of the quantum-mechanical state $|\Psi_{singlet}\rangle$. But in fact we can give a general proof that quantum-mechanical correlations cannot be used for signalling along the following lines.

Consider, quite generally, two systems A and B with associated Hilbert spaces $\mathbb{H}_A$ and $\mathbb{H}_B$. Let C be a third system which may interact in any way with system B, with associated Hilbert space $\mathbb{H}_C$. Denote $\mathbb{H}_B \otimes \mathbb{H}_C$ by $\overline{\mathbb{H}}_B$. a is any observable on $\mathbb{H}_A$, extended to $a \otimes I$ on $\mathbb{H}_A \otimes \overline{\mathbb{H}}_B$. Consider any state $|\Psi(t)\rangle$ of the triple system at time t and denote by $\text{Prob }(\lambda)_{a(t)\otimes I}^{|\Psi(t)\rangle}$ the probability that the time-dependent observable $a(t) \otimes I$ will yield the measurement result λ in the state $|\Psi(t)\rangle$. We shall work in the Dirac picture (see p. 12) so in the absence of perturbation $|\Psi(t)\rangle$ is constant, while in all cases $a(t) \otimes I$ evolves in time according to the unperturbed Hamiltonian.

Now perturb the system B in any way in the time interval (t, t') by the action of a unitary time-evolution operator $U_B(t', t)$ acting on $\overline{\mathbb{H}}_B$.

Then at time t', the state of the triple system is

$$|\Psi(t')\rangle = (I \otimes U_B(t', t))|\Psi(t)\rangle \tag{60}$$

We now compute

$$\text{Prob}(\lambda)_{a(t')\otimes I}^{|\Psi(t')\rangle} = \text{Prob }(\lambda)_{a'(t')\otimes I}^{|\Psi(t)\rangle} \tag{61}$$

where

$$\begin{aligned}a'(t') \otimes I &= (I \otimes U_B^{-1})(a(t') \otimes I)(I \otimes U_B) \\ &= (I \otimes U_B^{-1})(a(t') \otimes U_B) \\ &= a(t') \otimes (U_B^{-1} U_B) = a(t') \otimes I\end{aligned} \tag{62}$$

So

$$\text{Prob }(\lambda)_{a(t')\otimes I}^{|\Psi(t')\rangle} = \text{Prob }(\lambda)_{a(t')\otimes I}^{|\Psi(t)\rangle} \tag{63}$$

But the RHS of this equation is the probability of finding the result λ by measuring a on A at time t' in the absence of perturbation, since then the state vector at t' is the same as the state vector at t.

In other words, the probability of obtaining the result λ by measuring a on A at any time t' is independent of any possible perturbation of the system B between t and t'.

4.7. Summary of Conclusions

We have distinguished three basic approaches to the interpretation of QM, labelled A, B and C, and also five senses of locality: LOC_1,

LOC_2, LOC_3, LOC_4 and LOC_5. We can now answer the question 'Does QM predict a violation of locality?' in the form of a simple table showing which senses of locality can be violated, as indicated by a cross, in each of the three interpretations. A tick indicates that the corresponding sense of locality is not violated.

View	LOC_1	LOC_2	LOC_3	LOC_4	LOC_5
A	✓	✓	✗	✗	✓
B	✗	✓	✓	✓	✓
C	✓	✗	✓	✓	✓

Notes and References

Most of the material covered in this chapter is based on Redhead (1983).

The original proof of the Bell inequality was given in Bell (1964). This used the deterministic hidden-variable framework, and also assumed perfect anti-correlation when the two spin-meters measured parallel spin-components. These restrictions were relaxed in Bell (1971). See also Clauser and Horne (1974). An interesting alternative approach is provided in Wigner (1970). The Stapp–Eberhard approach to the Bell inequality discussed in Section 4.2. was initiated by Stapp (1971) and developed by Eberhard (1977). The mathematics, as distinct from the interpretation, of Eberhard's paper is used in Section 4.1, with improvements dues to Peres (1978) and Brody (1980). See also Peres and Zurek (1982).

For discussion of counterfactuals, see in particular Lewis (1973), Slote (1978), Lewis (1979), Bowie (1979), and for an approach that leads to the opposite conclusion to the one discussed in the text, see Thomason and Gupta (1980). A useful collection of reprints on this topic is given in Harper et al. (1981). Our own treatment is based on a simplified adaptation of Lewis (1973). For a critique of some of the assumptions involved in the proof of the Bell inequality even in the deterministic case, see in particular Fine (1974) and (1979), Brody and de la Peña-Auerbach (1979), and Brody (1980).

Stochastic hidden-variable theories and the relevance of factorizability to locality are discussed by Clauser and Horne (1974), Selleri and Tarozzi (1980), Fine (1981), Shimony (1981), Hellman (1982), Jarrett (1984) and Shimony (1984). The term 'passion-at-a-distance' is due to Shimony. The nonrobustness of the singlet state was first analysed in Redhead (1986). For the mark criterion for causal connectibility see Reichenbach (1928) and (1956) and Salmon (1984). The proof of the inequality (42) given in the text can be found in Clauser and Horne (1974), Appendix A. The fact that strict correlation combined with factorizability implies determinism was first pointed out by Suppes and Zanotti (1976).

Fine (1982a) and (1982b) has demonstrated an interesting mathematical property of the Bell inequality, viz. that it is a sufficient condition for the existence of joint distributions for all observables including incompatible ones. The fact that this does not contradict the claim made in this chapter that the Bell inequality can be derived without assuming joint distributions for incompatible observables has been argued in Redhead (1984) and Svetlichny *et al.* (1988). The basic point here is that Fine's theorem demonstrates the existence of joint distributions in a mathematical model of statistics that satisfy the Bell inequality. But this model may not reflect the frequencies arising in the real world.

Comprehensive reviews of the experiments designed to test the Bell inequality are provided by Clauser and Shimony (1978) and Pipkin (1978). The references for the more recent work of Aspect and his collaborators are Aspect, Grangier, and Roger (1981) and (1982), and Aspect, Dalibard, and Roger (1982). The most recent experiment of Perrie *et al.* is presented in their (1985). A critique of the auxiliary assumptions used in the Aspect experiment has been made in particular by Marshall *et al.* (1983). The Lo–Shimony experiment is discussed in great detail in Lo and Shimony (1981).

For the general proof that LOC_5 is not violated in QM see Eberhard (1978), Ghirardi *et al.* (1980), Page (1982) and Shimony (1984). A very simple and elegant proof is also provided by Jordan (1983). The proof given in the text follows Redhead (1986).

It can be shown that the QM violation of the Bell inequality never exceeds $2\sqrt{2}$. For an elegant demonstration see Landau (1986).

A recent general survey of the topics covered in this Chapter is provided by D'Espagnat (1984).

Bell's collected papers on the interpretation of QM are now available in Bell (1987).

5

The Kochen–Specker Paradox

THE interpretation of QM we described in Section 2.1 as view A has two parts to it:

1. It is possible to assign values to all observables in all states.
2. The statistics of these assigned values must coincide with the result of applying the statistical algorithm of QM.

The violation of the Bell inequality discussed in the preceding chapter is concerned with the impossibility of satisfying (2) subject to the constraints of locality. But so far we have assumed that (1) by itself poses no problems. The Kochen–Specker paradox is concerned with an algebraic contradiction that arises even at the preliminary stage (1).

How could such a problem arise? Suppose we were just asked to list some self-adjoint operators on a Hilbert space and then assign some real numbers as possessed values of the corresponding observables. The exercise would seem totally trivial. Thus we might fill in entries for $[Q]$ in Table 2.

Table 2. Value assignments for quantum-mechanical observables.

Operator	Observable	Value assignment $[Q]$
$\hat{Q}_1$	Q_1	α_1
$\hat{Q}_2$	Q_2	α_2
$\hat{Q}_3$	Q_3	α_3
.	.	.

Difficulties only arise if we introduce some restrictions on the possible values for $[Q]$. We have already introduced (p. 89) a Principle of Faithful Measurement (FM) which says that the result of measuring Q reveals the possessed value $[Q]$. FM motivates the following

$$\text{Value Rule (VR)}. \quad \text{Prob}(\lambda)_Q^{|\phi\rangle} = 0$$
$$\to [Q]^{|\phi\rangle} \neq \lambda$$

The idea behind VR is that if the QM probability for λ turning up on measurement is zero in the state $|\phi\rangle$ then λ cannot be one of the possessed values for Q in the state $|\phi\rangle$. Clearly VR is not a logical consequence of FM. In order to deduce VR from FM we need additional assumptions: (1) the identity of probability distributions for measured and possessed values; and (2) that, if there is zero probability for Q to possess the value λ, then $[Q]$ cannot equal λ. However, it would demonstrate a remarkable conspiracy to conceal possessed values on nature's part, if possessed values were always revealed on measurement, but repeated measurement failed to uncover with non-vanishing probability all those values which were possessed. If we reject such a conspiracy, then FM may indeed be regarded as justifying VR.

From VR we can easily obtain a result we know anyway from FM and the Quantization Algorithm.

The Spectrum Rule:

For observables with a discrete spectrum possible values for $[Q]$ are confined to the eigenvalues of the associated operator $\hat{Q}$.

But notice carefully that VR would make no sense for operators with a continuous spectrum, since it would imply that such an operator possesses no sharp value at all! In our later applications of VR we shall always be dealing with the finite-dimensional case, where VR does seem eminently reasonable.

Another consequence of VR is

The Eigenvector Rule: $[Q]^{|q_i\rangle} = q_i$

This has aleady been introduced in Chapter 3 as a consequence of the Einstein reality criterion (cf. Eq. 3.1). Here we have derived it from VR.

Although it is clear that VR provides a significant constraint on value assignments, it does not lead to any paradoxical contradictions. This unfortunate circumstance does occur, however, if, following Kochen and Specker (1967), we introduce a constraint on value assignments known as

The Kochen–Specker Paradox

The Functional Composition Principle (FUNC):
If $\hat{A}$ and $\hat{B}$ are two self-adjoint operators with associated observables A and B, and if there exists a function f such that $\hat{B} = f(\hat{A})$, then $[B]^{|\phi\rangle} = f([A]^{|\phi\rangle})$ for any state $|\phi\rangle$.

An alternative way of expressing FUNC is

$$[f(A)]^{|\phi\rangle} = f([A]^{|\phi\rangle}) \tag{1}$$

where $f(A)$ denotes the observable whose associated self-adjoint operator is $f(\hat{A})$.

Essentially, the idea of FUNC is that the algebraic structure of the operators should be mirrored in the algebraic structure of the possessed values of the observables. We shall consider the justification for FUNC a little later on, but first we shall show that, if FUNC constrains the value assignments, a contradiction will result, provided that the Hilbert space of state vectors has a dimension greater than two. This is known as the *Kochen–Specker Paradox*.

5.1. Demonstration of the Contradiction

We begin by deriving from FUNC a result we shall call

The Sum Rule:
If $\hat{A}$ and $\hat{B}$ commute then

$$[A+B]^{|\phi\rangle} = [A]^{|\phi\rangle} + [B]^{|\phi\rangle}$$

for an arbitrary state $|\phi\rangle$, where $A + B$ is the observable corresponding to the self-adjoint operator $\hat{A} + \hat{B}$.

Proof: We have already seen, in Section 1.3, that if $\hat{A}$ and $\hat{B}$ commute, then there exists a maximal operator $\hat{O}$ such that

$$\hat{A} = f(\hat{O})$$
$$\hat{B} = g(\hat{O})$$

for appropriate functions f and g.
Then $\hat{A} + \hat{B} = h(\hat{O})$
where $h = f + g$

So $[A+B]^{|\phi\rangle} = h([O]^{|\phi\rangle})$ using FUNC
$= f([O]^{|\phi\rangle}) + g([O]^{|\phi\rangle})$
$= [f(O)]^{|\phi\rangle} + [g(O)]^{|\phi\rangle}$ using FUNC
$= [A]^{|\phi\rangle} + [B]^{|\phi\rangle}$ QED.

It is interesting to notice that Fine and Teller (1977) have shown that FUNC as applied to observables with a discrete spectrum can itself be derived from a combination of the Sum Rule and the Value Rule. Also, by a simple adjustment of the above argument, we can extend the Sum Rule to the case of a denumerable infinity of summands. Consider now an arbitrary separable Hilbert space, and some maximal self-adjoint operator with a purely discrete spectrum of eigenvalues for which we can write the spectral expansion

$$\hat{Q} = q_1\hat{P}_1 + q_2\hat{P}_2 + q_3\hat{P}_3 + \text{-----} \quad (1)$$

We know that any of the projection operators $\hat{P}_k$, say, can be written as a function of Q

Thus $$\hat{P}_k = \chi_{q_k}(\hat{Q}) \quad (2)$$

where the characteristic function $\chi_{q_k}: \{q_i\} \to \{0, 1\}$ is defined by $\chi_{q_k}(q_i) = \delta_{ik}$.

In terms of observables, one way to measure P_k is to measure Q and apply the function χ_{q_k} to the result, i.e. P_k has the measured value 1 if and only if Q has the measured value q_k. But, from the discussion of Section 1.3, we know that this is not the only way of measuring P_k, since P_k can also be expressed as a function of other maximal observables which are incompatible with Q. For the moment, we are going to assume that all of these measured values correspond to one and the same possessed values $[P_k]^{|\phi\rangle}$ in the state $|\phi\rangle$ under consideration.

Now we know that $\{\hat{P}_i\}$ provide a resolution of the identity in the sense that

$$\hat{P}_1 + \hat{P}_2 + \hat{P}_3 + \ldots = \hat{I} \quad (3)$$

Applying the Sum Rule (with in general a denumerable infinity of summands) we get

$$[P_1]^{|\phi\rangle} + [P_2]^{|\phi\rangle} + [P_3]^{|\phi\rangle} + \ldots = [I]^{|\phi\rangle} \quad (4)$$

We now apply FUNC to tell us what the possible values of $[P_k]$ are:

Since $$\hat{P}_k^2 = \hat{P}_k$$

we have, by FUNC,

$$([P_k]^{|\phi\rangle})^2 = [P_k]^{|\phi\rangle}$$

So $$[P_k]^{|\phi\rangle}([P_k]^{|\phi\rangle} - 1) = 0$$

Therefore $$[P_k]^{|\phi\rangle} = 0 \text{ or } 1 \quad (5)$$

This result follows also from the Spectrum Rule, since the eigenvalues of $\hat{P}_k$ are just 0 and 1. But at the moment we are interested in seeing how far we can get using only FUNC. What is the value of $[I]^{|\phi\rangle}$? The Spectrum Rule tells us that it must be 1, the eigenvalue of $\hat{I}$. But again we can derive this formally from FUNC via

The Product Rule:

If $\hat{A}$ and $\hat{B}$ commute then

$$[AB]^{|\phi\rangle} = [A]^{|\phi\rangle} \cdot [B]^{|\phi\rangle}$$

where AB is the observable corresponding to the (self-adjoint) operator $\hat{A}\hat{B}$.

The proof of the Product Rule is immediately obtained along the same lines as we gave for the Sum Rule, so we shall not repeat the details.

Let us make the plausible assumption that for any state $|\phi\rangle$ we can always choose an observable Q such that $[Q]^{|\phi\rangle} \neq 0$. Then,

$$[Q]^{|\phi\rangle} = [I \cdot Q]^{|\phi\rangle} = [I]^{|\phi\rangle} \cdot [Q]^{|\phi\rangle}$$

Hence
$$[I]^{|\phi\rangle} = 1 \tag{6}$$

From (5) and (6) it is easily seen that (4) dictates the following constraint:

All but one of the numbers $[P_k]^{|\phi\rangle}$ are zero, the remaining one has the value 1.

This must apply to all possible sets of orthogonal projectors which provide a resolution of the identity in the Hilbert space. Since the projectors are in 1:1 correspondence with the rays of Hilbert space, the above constraint means that, for every complete orthogonal basis of unit vectors in the Hilbert space, we must be able to associate the number 1 with one vector and the number 0 with all the other vectors in the basis in a consistent manner.

It is convenient to translate this problem into a colouring problem for the surface of a unit hypersphere in Hilbert space. Can we colour the hypersphere with two colours red and blue, in such a way that the following conditions are satisfied?

1. Every point (unit vector) is coloured red or blue.
2. For every complete orthogonal set of unit vectors only one is coloured red.
3. Unit vectors belonging to the same ray have the same colour.

We now proceed to demonstrate what we shall call

The Two-Colour Theorem:

If the dimension of the Hilbert space is greater than two, the colouring of the unit hypersphere in the way described is *not* possible.

Let us denote the Two-Colour Theorem in a real (complex) Hilbert space of finite dimension N by $T_N^{\text{real(complex)}}$ and in an infinite-dimensional separable Hilbert space by $T_\infty^{\text{real(complex)}}$.

We notice at once that
$$T_N^{\text{real}} \to T_{N+1}^{\text{real}} \tag{7}$$
This follows by supposing the colouring is possible in the $(N+1)$-dimensional case (i.e. $\sim (T_{N+1}^{\text{real}})$) and considering the N-dimensional subspace orthogonal to any direction coloured blue. This will now itself induce a colouring of the unit hypersphere in the subspace in accordance with the specification of the theorem.

So
$$\sim (T_{N+1}^{\text{real}}) \to \sim (T_N^{\text{real}})$$
or, contrapositively, (7).

Repeated applications of (7) show that if we can prove T_N^{real} for any given N, then the theorem is true for any greater value of N; and indeed, T_∞^{real} will follow by a similar argument, in which, assuming the colouring is possible in the infinite-dimensional case, we show that it would be possible in a finite-dimensional subspace that includes a direction coloured red.

Furthermore, for any dimension, finite or infinite
$$T^{\text{real}} \to T^{\text{complex}} \tag{8}$$
since, if we could colour the unit hypersphere in the Hilbert space defined over a complex field, we could show the colouring to be possible for a real Hilbert space of the same dimension by considering some particular complete orthogonal set of vectors in the complex case, and generating a structure isomorphic to a real Hilbert space by considering all the orthogonal sets obtained by real orthogonal transformations from the initial set of orthonormal vectors.

With these results in mind, all we need to do is examine T_N^{real} for low values of N. We notice that T_2^{real} is false. This corresponds to colouring the unit circle in a real Euclidean plane. The construction for doing such a colouring is illustrated in Fig. 19. Alternate open-closed quadrants are coloured blue and red as indicated. It is then easily checked that, for any orthogonal pair of directions, one is indeed coloured red and the other blue, with opposite directions having the same colour.

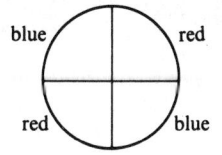

Fig. 19. Demonstration of the falsity of the Two-Colour Theorem in two dimensions.

We now show that T_3^{real} holds. First we shall give a plausibility argument due to Belinfante. Consider the unit sphere in Euclidean 3-space. Suppose the colouring had been effected. Then we would expect 1/3 of the surface of the sphere to be coloured red and 2/3 to be coloured blue, since for every orthogonal triad of directions one is coloured red and the remaining two blue. But every time we colour a point P, say, red, then we must colour the whole equator, with P as pole, blue, since any direction orthogonal to P gets coloured blue. This is illustrated in Fig. 20. PQR is one orthogonal triad of directions, so if P is red, then Q and R must be blue. Rotating this orthogonal triad about OP sweeps out the complete blue equator as shown. So it looks as though we can never end up with enough red points, if every red point is associated with an infinite number of blue points on the corresponding equator. The argument of course is quite unrigorous, since there is a lot of double counting if one remembers that each point on a blue equator lies on infinitely many other equators!

We now give a rigorous proof of T_3^{real}, using an ingenious argument of Kochen and Specker.

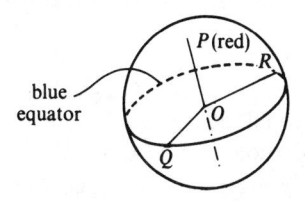

Fig. 20. Belinfante's proof of the Two-Colour Theorem for a real three-dimensional Hilbert space. O is the centre of the unit sphere, and P, Q, R are an orthogonal triad of directions. P is coloured red, and Q and R sweep out a blue equator as the triad is rotated about OP.

We first prove the following

Lemma:

There is a finite angular distance between any two points with opposite colour.

More specifically, we shall show that if 1 and 2 are any two points on the surface of the sphere with centre at O, and if we denote by θ_{12} the angle between the unit vectors $\overrightarrow{O1}$ and $\overrightarrow{O2}$, then, if $0 \leqslant \theta_{12} \leqslant \sin^{-1}(\frac{1}{3})$, the points 1 and 2 cannot be assigned the opposite colour.

Proof: We first introduce the representation of points and orthogonality relations on the sphere by means of a so-called Kochen–Specker diagram, in which points on the sphere are represented by vertices in the diagram; and if two points are in orthogonal directions, the corresponding vertices of the diagram are joined by a straight line. We show that the following Kochen–Specker diagram is constructible if $0 \leqslant \theta_{12} \leqslant \sin^{-1}(\frac{1}{3})$ (see Fig. 21).

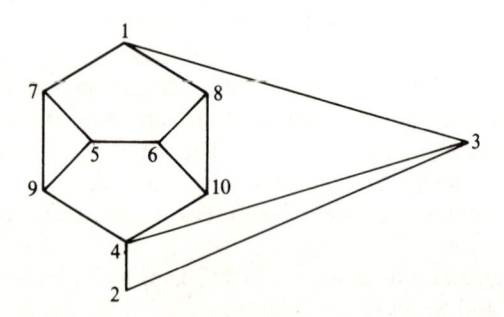

Fig. 21. Kochen–Specker diagram for ten points on the unit sphere in three-dimensional Euclidean space which is constructible if $0 \leqslant \theta_{12} \leqslant \sin^{-1}(1/3)$.

Suppose first that θ_{12} is any acute angle. Since 3 is orthogonal to 1 and 2, and 4 is orthogonal to 3, $\overrightarrow{O4}$ must be in the plane defined by $\overrightarrow{O1}$ and $\overrightarrow{O2}$. Since $\overrightarrow{O4}$ is orthogonal to $\overrightarrow{O2}$, we may choose 4 so that θ_{14} is also acute, and clearly $\theta_{14} = \pi/2 - \theta_{12}$, as sketched in Fig. 22.

Write $\overrightarrow{O5} = \mathbf{i}$, $\overrightarrow{O6} = \mathbf{k}$ and take a unit vector $\mathbf{j}$ orthogonal to $\mathbf{i}$ and $\mathbf{k}$ so as to complete a set of orthonormal vectors $\mathbf{i}, \mathbf{j}, \mathbf{k}$.

Then $\overrightarrow{O7}$, being orthogonal to $\mathbf{i}$, may be written as

$$\overrightarrow{O7} = (\mathbf{j} + x\mathbf{k})(1 + x^2)^{-\frac{1}{2}}$$

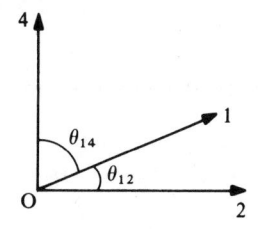

Fig. 22. Choice of angles θ_{14} and θ_{12} in the proof of the constructibility of the Kochen–Specker diagram shown in Fig. 21. O is the centre of the unit sphere.

Similarly,
$$\vec{O8} = (\mathbf{i} + y\mathbf{j})(1 + y^2)^{-\frac{1}{2}}$$
for some real numbers x and y.

The orthogonality relations in the diagram then force
$$\vec{O9} = (x\mathbf{j} - \mathbf{k})(1 + x^2)^{-\frac{1}{2}}$$
$$\vec{O10} = (y\mathbf{i} - \mathbf{j})(1 + y^2)^{-\frac{1}{2}}$$
But $\vec{O1}$ is orthogonal to $\vec{O7}$ and $\vec{O8}$, so we must have
$$\vec{O1} = (xy\mathbf{i} - x\mathbf{j} + \mathbf{k})(1 + x^2 + x^2y^2)^{-\frac{1}{2}}$$
Also $\vec{O4}$ is orthogonal to $\vec{O9}$ and $\vec{O10}$. So we must have
$$\vec{O4} = (\mathbf{i} + y\mathbf{j} + xy\mathbf{k})(1 + y^2 + x^2y^2)^{-\frac{1}{2}}$$
But then, taking the inner product of $\vec{O1}$ and $\vec{O4}$, we have at once
$$\cos\theta_{14} = \frac{xy}{((1 + x^2 + x^2y^2)(1 + y^2 + x^2y^2))^{\frac{1}{2}}}$$
and hence
$$\sin\theta_{12} = \frac{xy}{((1 + x^2 + x^2y^2)(1 + y^2 + x^2y^2))^{\frac{1}{2}}} \quad (9)$$

The expression (9) is easily seen to achieve a maximum value of 1/3 for $x = y = \pm 1$.

So the Kochen–Specker diagram is constructible if $0 \leq \theta_{12} \leq \sin^{-1}(1/3)$.

Now consider the colour map $V: S \to \{\text{red, blue}\}$, whose domain is the surface S of the unit sphere, and suppose that $V(1) = \text{red}$, and $V(2) = \text{blue}$, then from the Kochen–Specker diagram in Fig. 21, $V(1) = \text{red} \to V(7) = V(8) = V(3) = \text{blue}$.

But $V(2)$ = blue.
∴ $V(4)$ = red.
Hence $V(9) = V(10)$ = blue.
But $V(7) = V(9)$ = blue → $V(5)$ = red.
and $V(8) = V(10)$ = blue → $V(6)$ = red.
But $V(5) = V(6)$ = red is not possible, since 5 and 6 are orthogonal directions.

Hence we conclude that there is a minimum angular distance between points of opposite colour which is certainly greater than $\sin^{-1}(1/3)$, otherwise the above contradiction would be derivable.

There are two quite distinct ways in which we can use our lemma to prove T_3^{real}.

The first is based on an idea due to Bell (1966). In Section 1.5 we have already discussed effectively the colour map V, under the guise of a probability measure, with red identified with 1 and blue identified with zero. It was shown there that there is no minimum angular distance between points assigned opposite values (i.e. colours). So the argument now runs

$\sim (T_3^{\text{real}}) \to$ minimum angular distance between points assigned opposite colour.

But there can be no minimum angular distance between points assigned opposite colour.
∴ T_3^{real}.

The connection with Gleason's theorem should now be clear. If the colouring prohibited by T_3^{real} were possible, we could define a discontinuous probability measure on the projection lattice of a real Euclidean 3-space. But Gleason's theorem shows that there *are* no discontinuous measures for this case. Hence the colouring is not possible, in the sense that it contradicts the topology of the sphere (more specifically, the connectedness of the sphere as shown by the discussion already given in Section 1.5). So T_3^{real} follows as a simple corollary of Gleason's theorem.

We turn now to a second, quite distinct way of employing our lemma to prove the impossibility of the colouring considered in T_3^{real}. Suppose we take $\theta_{12} = 18° < \sin^{-1}(1/3)$, then we know that $V(1)$ = red → $V(2)$ = red. But now introduce four additional points, labelled 3, 4, 5, and 6, lying at 18° intervals along the equator through the points 1 and 2 as illustrated in Fig. 23. Then, repeating the above argument, $V(1)$ = red → $V(2)$ = red → $V(3)$ = red → $V(4)$ = red

The Kochen–Specker Paradox

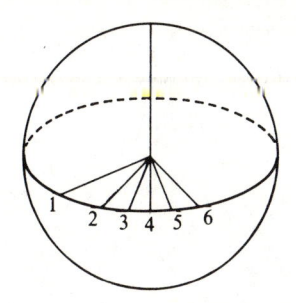

Fig. 23. 1, 2, 3, 4, 5, 6 are six points spaced at 18° intervals alongth an equator of the unit sphere. 1 and 6 are therefore orthogonal. It is shown in the text that if 1 is coloured red, then so also must 6.

$\rightarrow V(5) = $ red $\rightarrow V(6) = $ red. But $\theta_{16} = 5 \times 18° = 90°$. So, if any point on the sphere is coloured red, we have demonstrated two orthogonal red points, which contradicts the specification of the colouring. But, considering any three orthogonal points on the sphere, one of them must be coloured red. Hence the colouring is not possible, i.e. we have demonstrated T_3^{real}.

We can, in fact, put all this argument in the form of a Kochen–Specker diagram with a finite set of vertices which cannot be coloured in the specified way. All we have to do is to choose three arbitrary orthogonal points p_0, q_0, and r_0, and insert the appropriate fivefold repetition of the diagram shown in Fig. 21, with $\theta_{12} = 18°$, between each pair of vertices of the triangle $p_0 q_0 r_0$. The resulting beautiful example of an 'inconsistent' Kochen–Specker diagram, i.e. one which cannot be coloured, is shown in Fig. 24. Notice that, since a is orthogonal to r_0 and q_0, as also is p_0, then we can choose a to coincide with p_0. Similarly, we can identify the points b and q_0 and the points c and r_0. So the total number of vertices in the diagram is made up of eight vertices in each of the fifteen hexagons with three pairs of vertices identified, leaving $(8 \times 15) - 3 = 117$ distinct vertices.

In other words, what is demonstrated in Fig. 24 is a collection of 117 points on the unit sphere with the orthogonality relationships indicated, which can be constructed, but which cannot be coloured as specified. So we have again proved T_3^{real}. The above alternative method of proof is the one employed by Kochen and Specker themselves, who first produced the amazing 'cat's cradle' shown in Fig. 24. The advantage of their approach is that it deals with just a small finite number of observables, corresponding to the one-

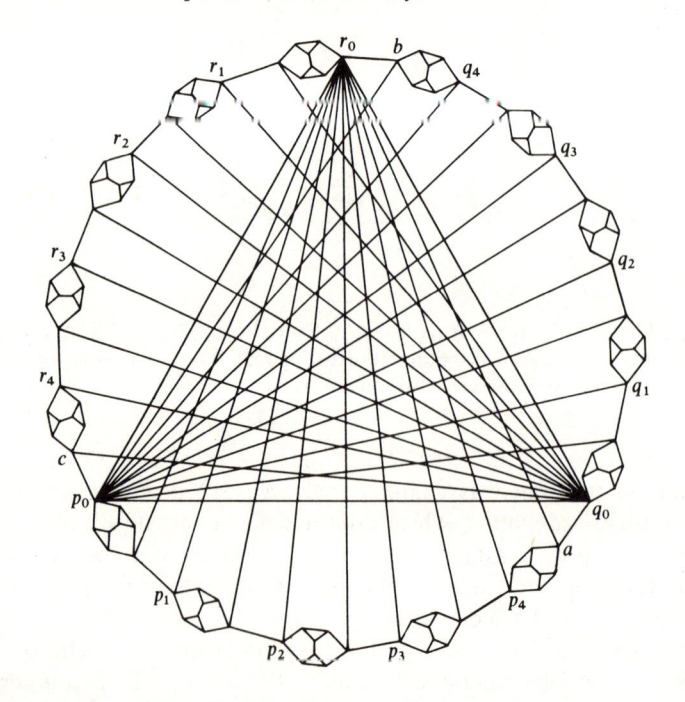

Fig. 24. Inconsistent Kochen–Specker diagram in three-dimensional Euclidean space.

dimensional projection operators in the ranges of which lie the 117 unit vectors specified in the Kochen–Specker construction. It should be remarked that, in their original paper, Kochen and Specker considered projection operators for a spin-1 system, described by a three-dimensional Hilbert space, with two-dimensional ranges rather than the one-dimensional projectors we have used. In fact, they consider the operators S_x^2, S_y^2, and S_z^2, corresponding to the square of the components of the angular momentum resolved along arbitrary orthogonal axes OX, OY, OZ, of the spin-1 system, for various orientations of these axes. Reference to Eq. (1.99) will show the connection between these operators and the one-dimensional projectors. Using S_x^2, S_y^2, and S_z^2, it is clear that the problem of assigning values reduces to solving the equation

$$[S_x^2]^{|\phi\rangle} + [S_y^2]^{|\phi\rangle} + [S_z^2]^{|\phi\rangle} = 2 \tag{10}$$

where each value assigned is again equal to 0 or 1.

Eq. (10) replaces Eq. (4) of our previous discussion. But the solution of (10) now requires that two of the three quantities $[S_x^2]^{|\phi\rangle}$, $[S_y^2]^{|\phi\rangle}$, $[S_z^2]^{|\phi\rangle}$ should have the value 1 and the remaining one the value 0, for any orientations of the axes OX, OY, and OZ. But this reduces to the same colouring problem as we considered before, if we make the trivial change of identifying red with 0 and blue with 1.

We have already shown how to express S_x^2, S_y^2, and S_z^2 as functions of the spin-Hamiltonian H_s introduced in Eq. (1.92). The explicit form of these functions is given in Eqs. (1.96), (1.97), and (1.98). These results provide an explicit method for measuring S_x^2, S_y^2, and S_z^2 in the case of an atom of orthohelium in its lowest triplet state. This is a spin-1 system, and Kochen and Specker remarked that the spin-Hamiltonian H_s is the perturbation arising when the atom is placed in a weak electric field of rhombic symmetry. So we can measure H_s by studying the change in energy levels when the field is applied. The measurement of S_x^2, S_y^2, and S_z^2 then consists of applying the functions specified in Eqs. (1.96), (1.97), and (1.98) to the measured result for H_s. For example, if the measurement result for H_s is $a + b$, then we infer that the values for S_x^2 and S_y^2 are each one, and the value of S_z^2 is zero, and so on.

Having demonstrated the Kochen–Specker paradox for any Hilbert space with dimension greater than two, we turn now to consider the status of FUNC, the constraint on value assignments which produced the contradiction.

5.2. The Justification of FUNC

We begin by considering the relation between FUNC and the theorem of QM we designated as STAT FUNC in Chapter 1 (see Eq. (1.48)). STAT FUNC, it will be recalled, states that

$$\text{Prob }(\Delta)_{f(Q)}^{|\phi\rangle} = \text{Prob }(f^{-1}(\Delta))_Q^{|\phi\rangle} \quad (11)$$

where we consider the case of a pure state $|\phi\rangle$.

Let us *reinterpret* this as a statement about long-run frequencies for *possessed* values of observables to lie in the indicated ranges in the given QM states $|\phi\rangle$. This move, as we have argued, is made plausible by FM. Can we replace (11) by

$$\text{Val }(\Delta)_{f(Q)}^{|\phi\rangle} = \text{Val }(f^{-1}(\Delta))_Q^{|\phi\rangle} \quad (12)$$

where Val denotes the truth valuation ascribed to propositions such as $(\Delta)_Q^{|\phi\rangle}$, this whole symbol now being used to denote the proposition that, on a particular occasion, the value of Q lies in the set Δ in the QM state $|\phi\rangle$?

On a realist construal of QM on the lines of view A, these truth values are, of course, always well-defined. More formally, Val. $\{\mathbb{P}\} \to \{T, F\}$ is a function that maps the set of propositions of the form $(\Delta)_Q^{|\phi\rangle}$ onto the two-element set of truth values, True and False. The function Val should be contrasted with the function Prob: $\{\mathbb{P}\} \to [0, 1]$, which maps the propositions $\mathbb{P}$ onto real numbers in the interval $[0, 1]$.

Now clearly, if (12) holds, then it would provide a very harmonious explanation of why (11) holds. (12) says Q has a value in $f^{-1}(\Delta)$ if, and only if, $f(Q)$ has a value in Δ. So the long-run frequencies, and hence the probabilities, must match because the outcomes match on each occasion. It is clear, however, that the converse implication does not hold. We cannot deduce (12) from (11). The match in long-run frequencies specified by (11) in no way guarantees a match in outcome on each particular occasion.

But (12) is actually equivalent to FUNC. To see, this we first notice that FUNC implies (12).

$$\text{Thus } [f(Q)]^{|\phi\rangle} \in \Delta$$
$$\text{iff } f([Q]^{|\phi\rangle}) \in \Delta$$
$$\text{iff } [Q]^{|\phi\rangle} \in f^{-1}(\Delta)$$

('iff' is a convenient abbreviation for 'if and only if') which is another way of expressing the content of (12). But also (12) implies FUNC. Thus

$$[Q]^{|\phi\rangle} \in \Delta \to [Q]^{|\phi\rangle} \in f^{-1}(f(\Delta)) \to [f(Q)]^{|\phi\rangle} \in f(\Delta)$$

Now take $\Delta = \{\lambda\}$, a singleton set. Then $[Q]^{|\phi\rangle} = \lambda \to [f(Q)]^{|\phi\rangle} = f(\lambda) = f([Q]^{|\phi\rangle})$, which is of course FUNC.

So the relationship between FUNC and STAT FUNC is the following. FUNC cannot be derived from STAT FUNC, which itself (understood in terms of the statistics of measurement results) is a theorem of the formalism of QM; but if FUNC were to hold it would make comprehensible why STAT FUNC holds. So in this sense STAT FUNC may be regarded as providing a plausibility argument for FUNC.

Let us turn now to consider how we might derive FUNC from other more fundamental principles. The fact that FUNC leads to contradiction would then show that one or other of these principles

The Kochen–Specker Paradox

had to be given up. We shall in fact show the following

$$\text{View } A \wedge Corr \wedge R \to \text{FUNC} \tag{13}$$

where, as usual, view A is our realist ascription of sharp values to all observables in all states, and *Corr* and *R* are the following principles:

Corr (Correspondence Rule):
There is a 1:1 correspondence between the set of self-adjoint operators and the set of observables.

R (Reality Principle):
If there is an operationally defined number associated with the self-adjoint operator $\hat{Q}$ (i.e. distributed probabilistically according to the statistical algorithm of QM for $\hat{Q}$), then there exists an element of reality (an observable) associated with that number and measured by it.

Be careful not to confuse this Reality Principle with the Einstein Reality Principle introduced in Chapter 3 and note that *Corr* assumes the absence of so-called superselection rules.

Let us now see how using these principles enables us to derive FUNC. Suppose we measure some observable Q in the QM state $|\phi\rangle$ which, on view A, and assuming faithful measurement FM, reveals the number $[Q]^{|\phi\rangle}$ possessed by the observable. Now form the number $f([Q]^{|\phi\rangle})$ for some arbitrary function f. This number is distributed probabilistically according to the statistical algorithm for $f(\hat{Q})$. This follows since

$$\text{Prob}(f([Q]^{|\phi\rangle}) \in \Delta) = \text{Prob}([Q]^{|\phi\rangle} \in f^{-1}(\Delta))$$
$$= \text{Prob}([f(Q)]^{|\phi\rangle} \in \Delta)$$

where we have employed STAT FUNC. Hence by R, there exists an observable associated with the self-adjoint operator $f(\hat{Q})$ and having the value $f([Q]^{|\phi\rangle})$. But, by *Corr*, there is only one such observable associated with $f(\hat{Q})$, which we denote as usual by $f(Q)$. So $[f(Q)]^{|\phi\rangle} = f([Q]^{|\phi\rangle})$, which is FUNC.

From (13) we see that there are basically three ways of escaping from the Kochen–Specker paradox:

1. We reject view A. $f(Q)$ does not possess a unique value, but is more properly a relational attribute, whose value depends on the context of measurement. If we envisage the situation $\hat{A} = f(\hat{B}) = g(\hat{C})$, where $[\hat{B}, \hat{C}] \neq 0$, then A can also be measured by measuring C and applying the function g to the result. There is no reason to suppose this number is the same as we would have got by measuring B, and applying the function f. So the value of A revealed by

measurement depends on the context of measurement, and expresses a holistic relational attribute involving the QM observable and the method or 'context' of measurement. This way of resolving the Kochen–Specker paradox was seen by Bell (1966) as an expression of the Bohrian insight into the nature of quantum-mechanical reality. The non-commutativity of $\hat{B}$ and $\hat{C}$ involved in the example reflects the incompatibility or mutual exclusiveness of the experimental arrangements.

We notice that this relationism need only apply to nonmaximal observables. Indeed, we know from the discussion in Chapter 1, p. 20, that if B and C are maximal and incompatible then A must be nonmaximal. But the proof of the Kochen–Specker paradox depends essentially on the constraint implied on the value assignments to incompatible maximal observables. Thus FUNC means that

$$[A]^{|\phi\rangle} = f([B]^{|\phi\rangle})$$

and also

$$[A]^{|\phi\rangle} = g([C]^{|\phi\rangle})$$

Hence $\qquad f([B]^{|\phi\rangle}) = g([C]^{|\phi\rangle}).$

Such constraints on value assignments to incompatible maximal observables only arise in the context of applying FUNC to value assignments for nonmaximal observables. Thus, in our derivation of the Kochen–Specker contradiction, the essential ingredient was that the values assigned to each one-dimensional projector (a nonmaximal observable) was the same independently of which set of orthogonal projectors it was considered to be a member, or equivalently of which maximal observable it was considered to be a function. If we were only concerned with value assignments to maximal observables which maintained functional relationships between them (the maximal observables), no contradiction would arise. This fact is plausible if we look at how the contradiction was arrived at; but a formal proof that no Kochen–Specker contradiction results if we consider only maximal observables was provided by Maczynski (1971).

2. However, we might respond by keeping view A, with its talk of possessed nonrelational attributes, and rejecting *Corr*. In other words, we now suppose that there are many different observables, corresponding in general to some particular operator $\hat{A}$. In order to deal with the Kochen–Specker paradox, it is sufficient to invoke this 'splitting' of observables for nonmaximal observables only, according to the discussion we have just given. So we now have the following

scheme. Corresponding to each nonmaximal operator there are many observables. Furthermore, these many observables are distinguished from one another by the functional relationship between their values and the values of observables corresponding to maximal operators. Thus, suppose as above, $\hat{A} = f(\hat{B})$, $\hat{A} = g(\hat{C})$, $[\hat{B}, \hat{C}] \neq 0$, where $\hat{B}$ and $\hat{C}$ are maximal so that we can unambiguously associate observables B and C with them. But $\hat{A}$ is nonmaximal, and with it we associate two observables—there may of course be more—A_B and A_C, and these observables are identified by their functional relations with B and C in respect of value assignments. Specifically

$$[A_B]^{|\phi\rangle} = f([B]^{|\phi\rangle}) \quad \text{and} \quad [A_C]^{|\phi\rangle} = g([C]^{|\phi\rangle}).$$

In a sense, observables corresponding to maximal operators are ontologically prior to those which correspond to nonmaximal operators. Knowing to which self-adjoint operator it corresponds is not sufficient to identify unambiguously a nonmaximal observable: we must know also to which maximal observable its values are related. It requires in this sense a context; and this splitting of nonmaximal observables, so that each nonmaximal operator now corresponds to many distinct observables, we shall refer to as *Ontological Contextuality*. This scheme for retaining a realism of possessed values (view A) but defeating the Kochen–Specker paradox, was proposed by van Fraassen (1973). In view of the profusion of physical magnitudes introduced, it was referred to by Clark Glymour as de-Ockhamizing QM ! The Kochen–Specker proof is blocked because there is no reason to think that

$$[A_B]^{|\phi\rangle} = [A_C]^{|\phi\rangle}$$

when $\hat{B}$ and $\hat{C}$ do not commute, which is what the Kochen–Specker argument requires in this contextualized terminology. Notice that in order to measure A_B, for example, we proceed by measuring B and applying the function f to the result. Similarly, A_C is measured by measuring C and applying the function g to the result. So, expressed in terms of measurement procedures, there is no difference from Bell's relational attribute approach discussed under (1). But ontologically the van Fraassen approach is quite different, with its realism of possessed values.

3. We might deny the Reality Principle R. To any nonmaximal operator $\hat{A}$ there corresponds a unique observable; but, assuming as before that $\hat{A} = f(\hat{B}) = g(\hat{C})$, then the unique observable A is not

necessarily measured by measuring B and applying f to the result. True, numbers obtained by such a procedure are distributed probabilistically according to the statistical algorithm for $\hat{A}$; but since we are denying R, these numbers do not necessarily measure any ontologically existing physical magnitude. The 'correct' method of measuring A might be to measure C and apply the function g to the result. But there is no reason to assume that this measurement of a 'real' A is equal in value to the number produced by the first procedure which, in the view we are propounding, does not measure anything real at all. This arguably is the approach of Arther Fine to avoiding the Kochen–Specker paradox. But the onus does seem to lie with a proponent of such a view to tell us which putative method of measuring A reveals the 'real' value, and which methods produce numbers which just 'hang in the air' and do not measure anything of ontological significance.

Since we are trying to see how far we can get in interpreting QM with a simple realism of possessed values, we shall adopt the second or van Fraassen way of dealing with the Kochen–Specker paradox, which deals with all measurement procedures on an equal basis, and reject the arbitrary distinction between 'genuine' and 'nongenuine' measurements entailed by the Fine approach.

There is, however, one obvious complication that needs discussion. Suppose we measure a nonmaximal observable directly without going through the intermediary of some maximal observable. Suppose, for example, in the angular momentum case, we measured the total angular momentum alone (a nonmaximal quantity), instead of a combined measurement of the total angular momentum together with the (compatible) component of angular momentum along some specified direction. This combined measurement would obviously be a maximal one. By altering the direction in which we measured the component of angular momentum, we would be specifying incompatible maximal measurements, which is the case we have already considered. But what of the nonmaximal measurement on its own? Now, for purposes of discussing how to deal with the Kochen–Specker paradox, we just do not *need* to answer this question. It would need answering only if we embarked on the enterprise of trying to fill out the realist interpretation, rather in the way the detailed hidden-variable theories do. For example, van Fraassen has proposed that nonmaximal measurements yield always one of the results that some maximal measurement would have yielded. The reader is referred to van Fraassen (1979) for further discussion.

We now want to point out that, in the van Fraassen split-observable approach, we do accept a restricted version of FUNC, call it FUNC*, which is relativized to a particular ontological context. Let us state FUNC* more carefully:

FUNC*:

Let $\hat{B}$ be a *maximal* self-adjoint operator and $\hat{A}$ and $\hat{D}$ self-adjoint operators, such that for functions h, f, and g we have the relations

$$\hat{A} = f(\hat{B}), \quad \hat{D} = g(\hat{B}), \quad \hat{A} = h(\hat{D})$$

then
$$[A_B]^{|\phi\rangle} = h([D_B]^{|\phi\rangle})$$

FUNC* follows at once from the defining relations for $[A_B]^{|\phi\rangle}$ and $[D_B]^{|\phi\rangle}$.

Thus $[A_B]^{|\phi\rangle} = f([B]^{|\phi\rangle}) = h \circ g([B]^{|\phi\rangle})$
$$= h([D_B]^{|\phi\rangle})$$

In particular, if $\hat{D}$ is also maximal, so we can identify D_B with the unique observable D,

$$[A_B]^{|\phi\rangle} = [A_D]^{|\phi\rangle}$$

So the value of A_B in a given state $|\phi\rangle$ does not depend on the maximal observable of which it is considered a function in the equivalence class of 1:1 functions of a given maximal observable. Thus, if we denote the equivalence class generated by B by $\{B\}$, our notation can be conveniently modified by defining a new symbol

$$[A]^{|\phi\rangle}_{\{B\}} \underset{Df}{=} [A_B]^{|\phi\rangle}$$

which serves to stress that $[A_B]^{|\phi\rangle}$ depends only on the equivalence class $\{B\}$, or (which comes to the same thing) the complete orthonormal basis of eigenvectors for $\hat{B}$ (in the case of the discrete spectrum we are considering for the purposes of this discussion).

With regard to our avowal of FUNC*, two questions immediately arise. Is it consistent with, and is it independent of, the Value Rule VR? Both questions can be answered in the affirmative. Let R be any maximal observable with a unique (up to phase) orthonormal basis of eigenvectors. Order the set in some way $|\phi_1\rangle, |\phi_2\rangle \ldots$. Then let n_0

be the first n such that $|\langle \phi_n|\phi\rangle|^2 \neq 0$ and define

$$[f(R)]_{\{R\}}^{|\phi\rangle} = f(\lambda_{n_0}) \qquad (14)$$

where λ_{n_0} is the eigenvalue corresponding to $|\psi_{n_0}\rangle$. By choice of n_0 this satisfies VR and also agrees with FUNC*. Hence we have established consistency. But now suppose (14) holds only if f is not the identity function, and define $[R]_{\{R\}}^{|\phi\rangle} = \lambda_{n_1}$ where n_1 is the second n such that $|\langle \phi_n|\phi\rangle|^2 \neq 0$, and λ_{n_1} is the eigenvalue corresponding to $|\phi_{n_1}\rangle$, then VR still holds but FUNC* is violated. Hence we have established independence.

On the question of whether FUNC* is consistent with the algebraic structure of *maximal* observables, we can cite the theorem of Maczynski, already referred to, to the effect that it is indeed possible to assign values to the set of maximal observables consistently with maintaining functional relationships between them.

Notes and References

The original presentation of the Kochen–Specker paradox is given in Kochen and Specker (1967).

An important discussion referred to in the text is Fine and Teller (1977). The Sum Rule and the Product Rule were introduced by Fine (1974). What we call the Two-Colour Theorem is extensively discussed by Belinfante (1973). For the justification of FUNC and its relation to STAT FUNC, see Healey (1979).

There has been considerable discussion on the question of whether the Sum Rule can be checked experimentally as a principle which supplements the QM formalism. The problem is examined in some detail in Redhead (1981) with a negative conclusion. The basic point at issue is that we cannot check $[A_B]^{|\phi\rangle} = [A_C]^{|\phi\rangle}$ if B and C cannot be simultaneously measured.

For ways out of the Kochen–Specker paradox referred to in the text, see Bell (1966), van Fraassen (1973) and (1979), and Fine (1974). FUNC* is introduced in Heywood and Redhead (1983). The independence and consistency of FUNC* with respect to VR were first pointed out by Fine.

For a proof of Maczynski's theorem, see Maczynski (1971).

6

Nonlocality and the Kochen–Specker Paradox

WITH regard to the proliferation of observables envisaged by the van Fraassen solution to the Kochen–Specker paradox, an interesting question arises in the case of two spatially separated systems. Is there a connection between ontological contextuality and nonlocality? This is the problem we shall investigate in this chapter.

6.1. Contextuality and Nonlocality

Let S_1 and S_2 be two spatially separated systems which may or may not have interacted in the past, and let H_1 and H_2 be their associated Hilbert spaces. As described in Section 1.6, the combined system $(S_1 + S_2)$ is associated with the tensor product space $\mathsf{H}_1 \otimes \mathsf{H}_2$. Let $\hat{A}$ be a maximal operator on the space H_1 and $\hat{B}$ a maximal operator on the space H_2. Now consider the observable $A \otimes I$ associated with the operator $\hat{A} \otimes \hat{I}$ for the joint system. This is not a maximal observable in the product space. We shall refer to it as a *locally maximal* observable, since we are assuming $\hat{A}$ is maximal on H_1.

Now the following problem arises. Since $A \otimes I$ is not maximal, should $[A \otimes I]^{|\phi\rangle}_{\{X\}}$ and $[A \otimes I]^{|\phi\rangle}_{\{Y\}}$, where X and Y are maximal incompatible observables in the product space, be treated as the values (in general unequal) of distinct 'split' observables? A straightforward application of ontological contextuality would lead us to do so. The intuitive consequences of doing this are peculiar. A sort of holism is involved. Observables maximal in the product space are somehow prior to observables maximal locally in the factor spaces. This may be called 'nonseparability', but we shall refer to it more specifically as failure of 'ontological locality'. Let us state the principle of Ontological Locality.

Ontological Locality (OLOC):

If H_1 and H_2 are the Hilbert spaces for two spatially separated systems and

$(\hat{A} \otimes \hat{I})$ is a locally maximal operator, then

$$[A \otimes I]\,|^\phi_{\{X\}}\rangle = [A \otimes I]\,|^\phi_{\{Y\}}\rangle$$

for any state $|\phi\rangle$ of the joint system where $\hat{X}$ and $\hat{Y}$ are both maximal operators on $\mathbb{H}_1 \otimes \mathbb{H}_2$ and $[\hat{X}, \hat{Y}] \neq 0$.

In other words, locally maximal observables on either of two spatially separated systems are not 'split' by ontological contextuality relative to the specification of different maximal observables for the joint system.

But suppose we could show that imposing OLOC on the joint system leads to a Kochen–Specker paradox. Then we would have given a purely algebraic proof that OLOC must be violated. Put another way, can we prove that Maczynski's theorem *cannot* be extended from maximal observables to include also locally maximal observables? This question, and with it the possibility of a purely algebraic proof of nonlocality, was raised by Bub in his (1976). Unfortunately, the question was finally answered in the negative. Maczynski's theorem could be extended to locally maximal observables. No Kochen–Specker contradiction could be derived requiring them to be split. This was shown by Demopoulos (1980). So we do not need to violate OLOC in order to rescue realism from inconsistency. It does not follow, of course, that OLOC is not in fact violated.

But how is OLOC related to the locality principles discussed in Chapters 3 and 4? In particular, in Chapter 4 we discussed LOC_3, the locality principle that one sharp value could not be changed into another sharp value by altering the setting of a remote piece of apparatus. To give this idea a more general setting, let us introduce the concept of *Environmental Contextuality*. This involves the idea that there is some nonquantum interaction between the system of interest and its surroundings, which occurs before the act of measurement and alters the values of observables of the system. These interactions are just what failure of LOC_3 must invoke in order to explain the violation of the Bell inequality. But presumably there is no reason why they should only occur immediately prior to a measurement. That is why we prefer to call this kind of contextuality 'environmental'. By their very nature we know next to nothing about these supposed interactions; but we do presume that, when they occur just before a measurement, they are in fact a nonquantum-mechanical

Nonlocality and the Kochen–Specker Paradox

interaction taking place between the measured system and the measuring apparatus, depending on (at the very least) the maximal observable on the measured system which the apparatus is set to measure, on the assumption that it is making some maximal measurement. Expanding our notation a little further, we write $[A]|_{\{B\}}^{\{\phi\}}(B)$ to indicate the value taken by the split observable A_B after the interaction between the system and an apparatus set to measure B but before the actual measurement takes place. The letter in parentheses labels the observable which the apparatus is set to measure. To make the meaning of this symbol clearer, consider $[A]|_{\{B\}}^{\{\phi\}}(C)$, where B and C are incompatible maximal observables. This simply means the value that the magnitude A_B would take if the measuring apparatus were set to measure C. Notice carefully that, although we are assuming $[A]|_{\{B\}}^{\{\phi\}}(C)$ to be a well-defined number, we can never find what it is by measurement, for we cannot measure A_B (via a measurement of B) and leave the apparatus set to measure C, since we are assuming B and C to be incompatible.

When we apply this idea of contextuality to spatially separated systems we come up with another form of locality which we shall call *Environmental Locality*.

Environmental Locality (ELOC):

If S_1 and S_2 are two spatially separated systems, Q an observable for S_1, and X and Y maximal observables for the joint system $(S_1 + S_2)$, then if the difference between an apparatus set to measure X and one set to measure Y is only in the setting of that part of it located at S_2, $[Q \otimes I]|_{\{X\}}^{\{\phi\}}(X) = [Q \otimes I]|_{\{X\}}^{\{\phi\}}(Y)$

In other words, the value possessed by a local observable cannot be changed by altering the arrangement of a remote piece of apparatus which forms part of the measurement context for the combined system.

Note that we have not presumed OLOC in the specification of ELOC. In this sense it is a generalization of the idea incorporated in LOC_3 which involved a tacit assumption of OLOC. In fact, as we shall discuss later (Section 6.4), it is only when OLOC obtains that ELOC can properly be called a locality principle at all.

What we shall now do is to link these two notions of locality in the following way. We shall show for an appropriate physical system that

$$VR \wedge FUNC^* \wedge ELOC \wedge OLOC \rightarrow \text{Contradiction}$$

where VR is the Value Rule introduced on p. 120 above, and FUNC* is the noncontroversial version of FUNC defined on p. 137. The contradiction is derived from a connection between value assignments that is very closely connected with FUNC, although, as we shall see, it differs from FUNC in a rather subtle way.

Hence

$$VR \wedge FUNC^* \to (\sim ELOC) \vee (\sim OLOC)$$

So we are presented with a dilemma. If we hold on to VR and FUNC*, then we must violate either ELOC or OLOC (or both). The implications of grasping either horn of this dilemma will be discussed in Section 6.4. We turn now to the formal proof of our main result.

6.2. The Comeasurable Value Rule

We begin by slightly amending VR to take proper account of environmental contextuality. Considering the case of a maximal observable R, we write

$$\text{Prob}(\lambda)_R^{|\phi\rangle} = 0 \to [R]^{|\phi\rangle}(R) \neq \lambda$$

i.e., the vanishing probability of observing measurement results constrains the values R may have in the measurement context of R, and is silent about the quantity $[R]^{|\phi\rangle}(P)$, for example, where P is an incompatible maximal observable. Similarly, we adapt FUNC* to take account of environmental contextuality: for any environmental context C

$$[A]_{\{B\}}(C) = h([D]_{\{B\}}^{|\phi\rangle}(C))$$

where $\hat{B}$ is maximal and $\hat{A} = h(\hat{D})$, $\hat{A} = f(\hat{B})$ and $\hat{D} = g(\hat{B})$

In particular we have the result

$$[A]_{\{B\}}^{|\phi\rangle}(B) = f([B]^{|\phi\rangle}(B))$$

which is the form in which we shall employ FUNC* later in this section.

From VR and FUNC* we shall now derive what we call the Comeasurable Value Rule (CVR). Fine (1974) moots, but quickly rejects, a constraint on value assignments which we call the Extended Value Rule. It is as follows:

The Extended Value Rule (EVR):

If $\hat{Q}_1$ and $\hat{Q}_2$ commute, $\text{Prob}(\lambda, \mu)_{Q_1, Q_2}^{|\phi\rangle} = 0 \to$ either $[Q_1]^{|\phi\rangle} \neq \lambda$ or $[Q_2]^{|\phi\rangle} \neq \mu$, where $\text{Prob}(\lambda, \mu)_{Q_1, Q_2}^{|\phi\rangle}$ denotes the quantum-mechanical joint

Nonlocality and the Kochen–Specker Paradox

probability of finding measurement results λ for the observable Q_1 and μ for the observable Q_2 in the state $|\phi\rangle$.

The reason Fine so swiftly rejected this rule is because it is easily shown to imply FUNC. Thus we have, from the statistical algorithm of QM, if $\hat{W}_\phi$ is the density operator associated with the state $|\phi\rangle$ and χ_Δ denotes the usual characteristic function associated with the set Δ,

$$\text{Prob}(\lambda, \mu)|^{|\phi\rangle}_{Q, f(Q)} = \text{Tr}(\hat{W}_\phi \cdot \chi_\lambda(\hat{Q}) \cdot \chi_\mu(f(\hat{Q})))$$
$$= \text{Tr}(\hat{W}_\phi \cdot \chi_\lambda(\hat{Q}) \cdot \chi_{f^{-1}(\mu)}(\hat{Q}))$$
$$= \text{Tr}(\hat{W}_\phi \cdot \chi_{\{\lambda\} \cap f^{-1}(\mu)}(\hat{Q})).$$

Hence, if $\lambda \notin f^{-1}(\mu)$, i.e., if $\mu \neq f(\lambda)$, it follows that

$$\text{Prob}(\lambda, \mu)|^{|\phi\rangle}_{Q, f(Q)} = 0$$

and hence by EVR

$$[Q]^{|\phi\rangle} \neq \lambda \quad \text{or} \quad [f(Q)]^{|\phi\rangle} \neq \mu$$

But suppose

$$[Q]^{|\phi\rangle} = \lambda$$

then it follows that

$$[f(Q)]^{|\phi\rangle} \neq \mu, \forall \mu \neq f(\lambda)$$

i.e.

$$[f(Q)]^{|\phi\rangle} = f(\lambda) = f([Q]^{|\phi\rangle})$$

which is FUNC.

We shall now introduce our CVR as a restriction of EVR to certain comeasurable observables. It is a necessary condition for two observables to be comeasurable that their associated self-adjoint operators commute, but this is not sufficient when account is taken of contextuality. It is certainly a sufficient condition for genuine comeasurability that the two observables Q_1 and Q_2 are defined in the context of the same maximal observable R.

We now introduce our new rule:

The Comeasurable Value Rule (CVR):

If $\hat{Q}_1$ and $\hat{Q}_2$ commute, and $\hat{R}$ is a maximal operator such that $\hat{Q}_1 = f(\hat{R})$ and $\hat{Q}_2 = g(\hat{R})$ for functions f and g, then $\text{Prob}(\lambda, \mu)|^{|\phi\rangle}_{Q_1, Q_2} = 0 \rightarrow$ either $[Q_1]^{|\phi\rangle}_{\{R\}}(R) \neq \lambda$ or $[Q_2]^{|\phi\rangle}_{\{R\}} \neq \mu$.

We shall now show that CVR follows from VR and FUNC*. It is easy to see that

$$\text{Prob}\,(\lambda,\mu)^{|\psi\rangle}_{Q_1,Q_2} = \text{Tr}\,(\hat{W}_\psi \cdot \chi_\lambda(f(\hat{R})) \cdot \chi_\mu(g(\hat{R})))$$
$$= \text{Tr}\,(\hat{W}_\phi \cdot \chi_{f^{-1}(\lambda)}(\hat{R}) \cdot \chi_{g^{-1}(\mu)}(\hat{R}))$$
$$= \text{Tr}\,(\hat{W}_\phi \cdot \chi_{f^{-1}(\lambda) \cap g^{-1}(\mu)}(\hat{R}))$$
$$= \text{Prob}\,(f^{-1}(\lambda) \cap g^{-1}(\mu))^{|\phi\rangle}_R$$

Hence from the Value Rule

$$\text{Prob}\,(\lambda,\mu)^{|\phi\rangle}_{Q_1,Q_2} = 0 \to [R]^{|\phi\rangle}(R) \notin f^{-1}(\lambda) \cap g^{-1}(\mu)$$

But suppose

$$[Q_1]^{|\phi\rangle}_{\{R\}}(R) = \lambda \quad \text{and} \quad [Q_2]^{|\phi\rangle}_{\{R\}}(R) = \mu$$

Then by FUNC*

$$[R]^{|\phi\rangle}(R) \in f^{-1}(\lambda) \quad \text{and} \quad [R]^{|\phi\rangle}(R) \in g^{-1}(\mu)$$

Thus

$$[R]^{|\phi\rangle}(R) \in f^{-1}(\lambda) \cap g^{-1}(\mu)$$

Hence

$$\text{Prob}\,(\lambda,\mu)^{|\phi\rangle}_{Q_1,Q_2} = 0 \to \sim([Q_1]^{|\phi\rangle}_{\{R\}}(R) = \lambda \quad \text{and} \quad [Q_2]^{|\phi\rangle}_{\{R\}}(R) = \mu)$$
$$\to \text{either}\,[Q_1]^{|\phi\rangle}_{\{R\}}(R) \neq \lambda \quad \text{or} \quad [Q_2]^{|\phi\rangle}_{\{R\}}(R) \neq \mu$$

which is just our CVR. Conversely, from CVR it is clear that we can derive only the restricted, acceptable version of FUNC, viz. FUNC*.

Let us develop a particular case of CVR which will interest us especially. Suppose $\hat{Q} \otimes \hat{I}$ and $\hat{I} \otimes \hat{Q}'$ are self-adjoint operators on some product space $\mathbb{H}_1 \otimes \mathbb{H}_2$, describing the states of two spatially separated systems S_1 and S_2. We use the convention that, in the tensor product $\hat{A} \otimes \hat{B}$ of two operators, the left-hand operator $\hat{A}$ acts on the space $\mathbb{H}_1$ and the right-hand operator $\hat{B}$ on the space $\mathbb{H}_2$. Let $\hat{Q} = h(\hat{A})$ and $\hat{Q}' = k(\hat{B})$ for functions h and k. Then $\hat{Q} \otimes \hat{I} = h(\hat{A}) \otimes \hat{I} = h(\hat{A} \otimes \hat{I})$. Similarly $\hat{I} \otimes \hat{Q}' = k(\hat{I} \otimes \hat{B})$. We suppose $\hat{A}$ and $\hat{B}$ are nondegenerate on their respective components of the product space, so $A \otimes I$ and $I \otimes B$ are locally maximal observables. Let $\hat{A} = \sum_i \alpha_i \hat{P}_i$ and $\hat{B} = \sum_i \beta_i \hat{P}'_i$ be the spectral resolutions of $\hat{A}$ and $\hat{B}$. Now any observable with associated self-adjoint operator $\hat{O} = \sum_{ij} c_{ij} \hat{P}_i \otimes \hat{P}'_j$, where $c_{ij} = F(\alpha_i, \beta_j)$ and $F: \mathbb{R}^2 \to \mathbb{R}$ is 1-1, is sufficient to show that CVR will apply to $\hat{Q} \otimes \hat{I}$ and $\hat{I} \otimes \hat{Q}'$. Since F is 1-1 there are functions f and g such that $f(c_{ij}) = \alpha_i$ and $g(c_{ij}) = \beta_j$, so

$$hf(\hat{O}) = h\left(\sum_{ij} f(c_{ij}) \hat{P}_i \otimes \hat{P}'_j\right)$$
$$= h\left(\sum_i \alpha_i \hat{P}_i \otimes \sum_j \hat{P}'_j\right)$$
$$= h(\hat{A} \otimes \hat{I})$$
$$= \hat{Q} \otimes \hat{I}$$

Similarly

$$kg(\hat{O}) = \hat{I} \otimes \hat{Q}'$$

Thus, in a product space $\mathbb{H}_1 \otimes \mathbb{H}_2$

$$\text{Prob}(X, Y)^{|\phi\rangle}_{\hat{Q} \otimes I, I \otimes Q'} = 0$$

by CVR implies

either $[Q \otimes I]^{|\phi\rangle}_{\{O\}}(O) \neq X$ or $[I \otimes Q']^{|\phi\rangle}_{\{O\}}(O) \neq Y$ (1)

Before applying this result, we want to give a succinct formulation of ELOC and OLOC. To this end we first define the symbol $\langle A, B \rangle$ to mean just the maximal observable O for the joint system whose associated operator $\hat{O}$ is constructed from the operators $\hat{A}$ and $\hat{B}$ associated with the component systems in the way described. Note that the bracket symbol denotes a function defined on the ordered pair (A, B) which is (partly) specified by the function F.

Now, in order to measure O we connect up two pieces of apparatus, one interacting with S_1 and adjusted to measure A on S_1, the other interacting with S_2 and adjusted to measure B on S_2. The ordered pair of these measurement results is then subjected to the function F. The resulting number is a measurement of O. The environmental context referring to a measurement of O can thus be spelled out as the ordered pair (A, B), indicating that the apparatus interacting with S_1 is set to measure A and the apparatus interacting with S_2 is set to measure B.

Our result (1) can now be expressed in the following form:

If $\quad\quad\quad\quad \text{Prob}(X, Y)^{|\phi\rangle}_{Q\otimes I, I\otimes Q'} = 0$

then $\quad\quad$ either $[Q\otimes I]^{|\phi\rangle}_{\{\langle A, B\rangle\}} (A, B) \neq X$ or

$$[I\otimes Q']^{|\phi\rangle}_{\{\langle A, B\rangle\}} (A, B) \neq Y \quad\quad\quad (2)$$

The suffix $\{\langle A, B\rangle\}$ shows that it does not matter which 1-1 function we choose for F in the specification of $\langle A, B\rangle$. We shall follow the convention that, in an ordered pair of symbols, the first member always refers to system S_1 and the second member to system S_2.

We are now in a position to state our two locality principles (one of them, OLOC, in a slightly extended form) using our new notation. For all Q, A, B, C, D, and E, where $Q = h(A)$ for some function h, and A, B, C, D, and E are all maximal

OLOC:
$$[Q\otimes I]^{|\phi\rangle}_{\{\langle A, B\rangle\}} (D, E) = [Q\otimes I]^{|\phi\rangle}_{\{\langle A, C\rangle\}} (D, E) \quad\quad (3)$$

ELOC:
$$[Q\otimes I]^{|\phi\rangle}_{\{\langle A, B\rangle\}} (D, E) = [Q\otimes I]^{|\phi\rangle}_{\{\langle A, B\rangle\}} (D, C) \quad\quad (4)$$

Notice that in the formulation (3) of OLOC we have employed FUNC* to write

$$[Q\otimes I]^{|\phi\rangle}_{\{\langle A, B\rangle\}} (D, E) = h([A\otimes I]^{|\phi\rangle}_{\{\langle A, B\rangle\}} (D, E))$$

and

$$[Q\otimes I]^{|\phi\rangle}_{\{\langle A, C\rangle\}} (D, E) = h([A\otimes I]^{|\phi\rangle}_{\{\langle A, C\rangle\}} (D, E))$$

Eq. (3) then follows from our previous formulation of OLOC in terms of locally maximal observables such as $A\otimes I$. *Mutatis mutandis*, we can apply (3) and (4) to observables $I\otimes Q'$ which are local for S_2.

6.3. The Incompatibility of CVR and Locality

Now we shall employ the results (2), (3), and (4) to derive a contradiction. In the proof, we shall again consider as one system, $S_1 + S_2$, two spatially separated systems S_1 and S_2. We suppose for simplicity that each system is associated with a Hilbert space of N-dimension, and consider two locally maximal self-adjoint operators $\hat{A} \times \hat{I}$ and $\hat{I} \times \hat{B}$. Let $\hat{A}$ have N distinct eigenvalues written in some arbitrary order as $a_1, \ldots, a_N$, and $\hat{B}$ have N distinct eigenvalues written in some arbitrary order as $b_1, \ldots, b_N$. Let the state of the combined system be

$$|\Psi\rangle = \sum_{m=1}^{N} C_m |a_m\rangle |b_m\rangle \qquad (5)$$

where C_m are unspecified complex coefficients, $|a_m\rangle$ denotes the eigenvector of $\hat{A}$ associated with the eigenvalue a_m, and $|b_m\rangle$ denotes the eigenvector of $\hat{B}$ associated with the eigenvalue b_m. $|\Psi\rangle$ is thus a linear combination of simultaneous eigenvectors of $\hat{A} \otimes \hat{I}$ and $\hat{I} \otimes \hat{B}$, the mth eigenvalue of $\hat{A}$ being correlated with the mth eigenvalue of $\hat{B}$.

Now consider some nonmaximal self-adjoint operator $\hat{Q}$ such that $\hat{Q} = f(\hat{A}) = g(\hat{A}')$ for suitable functions f and g, where $\hat{A}'$ is another maximal operator which does not commute with $\hat{A}$. By construction we have the result

$$\text{Prob}\,(a_m, y)_{A\otimes I, I\otimes B}^{|\Psi\rangle} = 0, \ \forall y \neq b_m \qquad (6)$$

Hence, applying (2), we obtain

$$[A\otimes I]_{\{\langle A,B\rangle\}}^{|\Psi\rangle}(A,B) = a_m$$
$$\to [I\otimes B]_{\{\langle A,B\rangle\}}^{|\Psi\rangle}(A,B) \neq y, \ \forall y \neq b_m$$
$$\to [I\otimes B]_{\{\langle A,B\rangle\}}^{|\Psi\rangle}(A,B) = b_m \qquad (7)$$

Also

$$\text{Prob}\,(f(x), b_m)_{f(A\otimes I), I\otimes B}^{|\Psi\rangle} = \text{Prob}\,(f^{-1}(f(x)), b_m)_{A\otimes I, I\otimes B}^{|\Psi\rangle}$$

by an obvious extension of our notation to allow for the value of $A\otimes I$ revealed by measurement to be a member of the set $f^{-1}(f(x))$. But

$$\text{Prob}\,(f^{-1}(f(x)), b_m)_{A\otimes I, I\otimes B}^{|\Psi\rangle} = 0$$
$$\forall x \text{ such that } a_m \notin f^{-1}(f(x))$$
i.e. such that $f(a_m) \neq f(x)$.

Hence, again using (2)

$$[I\otimes B]_{\{\langle A',B\rangle\}}^{|\Psi\rangle}(A',B) = b_m \to [f(A\otimes I)]_{\{\langle A',B\rangle\}}^{|\Psi\rangle}(A',B)$$
$$\neq f(x), \ \forall x \text{ such that } f(a_m) \neq f(x)$$
$$\to [f(A\otimes I)]_{\{\langle A',B\rangle\}}^{|\Psi\rangle}(A',B) = f(a_m) \qquad (8)$$

We now apply OLOC and ELOC in the form (3) and (4), suitably transposed to system S_2, to give

$$[I\otimes B]_{\{\langle A,B\rangle\}}^{|\Psi\rangle}(A,B) = [I\otimes B]_{\{\langle A',B\rangle\}}^{|\Psi\rangle}(A',B)$$

So from (7) and (8) we obtain

$$[A \otimes I]^{|\Psi\rangle}_{\{\langle A,B\rangle\}} (A, B) = a_m \rightarrow [f(A \otimes I]^{|\Psi\rangle}_{\{\langle A',B\rangle\}} (A', B) = f(a_m) \quad (9)$$

or, more succinctly

$$[f(A \otimes I]^{|\Psi\rangle}_{\{\langle A',B\rangle\}} (A', B) = f([A \otimes I]^{|\Psi\rangle}_{\{\langle A,B\rangle\}} (A, B)) \quad (10)$$

Now (10) is not quite FUNC, since the latter principle preserves functional relationships between the coexisting values attributed to observables, whereas in (10) we have different environmental contexts on the two sides of the equation. Nevertheless, (10) can be used to demonstrate a contradiction just as well as the noncontextualized form of FUNC. To see this, let us specialize to the case of two spin-1 systems, prepared so that the combined system $S_1 + S_2$ is in the singlet state of the total spin. This example has already been treated in Section 1.8.

In terms of the eigenstates of the Kochen–Specker spin-Hamiltonian

$$\hat{H}_s = a\hat{S}_x^2 + b\hat{S}_y^2 + c\hat{S}_z^2 \quad (11)$$

with $a, b,$ and c distinct real numbers, the singlet eigenstate of the total spin is given by (cf. Eq. (1.108))

$$|\Psi_{\text{singlet}}\rangle = \frac{1}{\sqrt{3}}(|a+c\rangle|a+c\rangle - |b+c\rangle|b+c\rangle - |a+b\rangle|a+b\rangle) \quad (12)$$

This is of the form (5) to which our general analysis applies. Take now for $\hat{A}'$ the operator

$$\hat{H}'_s = a\hat{S}_{x'}^2 + b\hat{S}_{y'}^2 + c\hat{S}_z^2 \quad (13)$$

where the new set of orthogonal directions denoted by the labels X', Y', and Z are obtained from those denoted by X, Y, Z by a rotation about the Z-axis. Finally, take for Q the operator

$$\hat{S}_z^2 = f(\hat{H}_s) = f(\hat{H}'_s) \quad (14)$$

where $f: \mathbb{R} \rightarrow \mathbb{R}$ is defined by (cf. Eq. (1.98))

$$f(x) = (c-a)^{-1}(b-c)^{-1}(x-(a+b))(x-2c) \quad (15)$$

(Note that, since f does not possess an inverse, (14) does not imply $H_s = H'_s$!) Since $f(\hat{H}_s \otimes \hat{I}) = f(\hat{H}_s) \otimes \hat{I} = \hat{S}_z^2 \otimes \hat{I}$, our result (10) can

Nonlocality and the Kochen–Specker Paradox

be expressed in an abbreviated notation in the form

$$[S_z^2 \otimes I]_{\{H_s' \otimes I\}}^{|\Psi_{\text{singlet}}\rangle}(H_s' \otimes I) = f([H_s \otimes I]^{|\Psi_{\text{singlet}}\rangle}(H_s \otimes I)) \quad (16)$$

where we have suppressed any contextuality parameter, either ontological or environmental, on which the values of the observables do not depend in the light of OLOC and ELOC. (We write the contextuality parameters as $H_s' \otimes I$ and $H_s \otimes I$ to indicate, in accordance with our convention for the order of factors in a tensor product, which system we are referring to. Having dropped the ordered pair notation used in (10), we need some other method of distinguishing which system H_s or H_s' refers to.)

Now (16) assigns to $[S_z^2 \otimes I]_{\{H_s' \otimes I\}}^{|\Psi_{\text{singlet}}\rangle}(H_s' \otimes I)$ and, most importantly, to the direction labelled Z, a number with the following two properties:

1. It is independent of the orientation of the X'- and Y'-axes used in specifying $\hat{H}'$.
2. It has the value 0 or 1.

Condition (2) follows from the fact that $H_s \otimes I$ must be assigned one of its eigenvalues, and the action of f as specified in (15) is to project the set of eigenvalues onto the set $\{0, 1\}$. Now we can repeat the argument which led to (16), using instead an orthogonal triad X', Y, and Z', so that we find that

$$[S_y^2 \otimes I]_{\{H_s'' \otimes I\}}^{|\Psi_{\text{singlet}}\rangle}(H_s'' \otimes I)$$

and hence the direction Y is assigned some number which is a different function, say g, of $[H_s \otimes I]^{|\Psi_{\text{singlet}}\rangle}(H_s \otimes I)$, where

$$\hat{H}_s'' = a\hat{S}_{x'}^2 + b\hat{S}_y^2 + c\hat{S}_{z'}^2$$

This number again has the properties:

1. It is independent of the X'- and Z'-axes used in specifying $\hat{H}_s''$.
2. It has the value 0 or 1.

The function g is given by (cf. Eq. (1.97))

$$g(x) = (b-c)^{-1}(a-b)^{-1}(x-(c+a))(x-2b) \quad (17)$$

Finally, the argument is repeated for $S_x^2 \otimes I$ and the number assigned to $[S_x^2 \otimes I]_{\{H_s''' \otimes I\}}^{|\Psi_{\text{singlet}}\rangle}(H_s''' \otimes I)$, and hence the direction X, is a new function, say h, of $[H_s \otimes I]^{|\Psi_{\text{singlet}}\rangle}(H_s \otimes I)$, with the properties:

1. It is independent of the Y'- and Z'-axes used in specifying a rotated operator
$$H_s''' = a\hat{S}_{\gamma'}^2 + b\hat{S}_{y'}^2 + c\hat{S}_{z'}^2.$$

2. It has the value 0 or 1.

The function h is given by (cf. Eq. (1.96))
$$h(x) = (a-b)^{-1}(c-a)^{-1}(x-(b+c))(x-2a) \qquad (18)$$

But the three functions f, g, and h have the property that, acting on any eigenvalue of $\hat{H}_s \otimes \hat{I}$, the sum of the three values is always 2. This can be checked at once from equations (15), (17), and (18), with x given the value $a+b$, $a+c$, or $b+c$.

So our final result is to assign to three arbitrarily chosen orthogonal directions X, Y, Z three unique numbers, each of which is 0 or 1, and whose sum is 2. But such an assignment of numbers is known to be impossible for an appropriately chosen finite set of orthogonal triads of directions in Euclidean 3-space. This is what Kochen and Specker showed explicitly.

But notice that the numbers $[S_z^2 \otimes I]_{\{H_s' \otimes I\}}^{|\Psi_{\text{singlet}}\rangle}(H_s' \otimes I)$,

$[S_y^2 \otimes I]_{\{H_s'' \otimes I\}}^{|\Psi_{\text{singlet}}\rangle}(H_s'' \otimes I)$ and $[S_x^2 \otimes I]_{\{H_s''' \otimes I\}}^{|\Psi_{\text{singlet}}\rangle}(H_s''' \otimes I)$

we use to get the contradiction, not only cannot in general be measured simultaneously, since $\hat{H}_s'$, $\hat{H}_s''$, and $\hat{H}_s'''$ do not commute, but cannot even be said to coexist simultaneously, owing to the differing environmental contexts. In contrast with the original Kochen–Specker paradox, we are dealing now not with simultaneously existing value assignments (albeit not simultaneously measurable) but with numbers which *would* be assigned on the assumption of differing environmental contexts.

6.4. Implications

The upshot of our argument is that a quantum realist who wishes to impose the Value Rule, and does not want to deny the innocent-looking FUNC*, is bound to accept some form of nonlocality—that is, he must deny either OLOC or ELOC.

Let us look at the two horns of the dilemma in turn. Violating OLOC means that we can no longer specify the properties of one system independently of a specification of properties relating to the whole combined system. This leads to an ontological holism, in which it is impossible to make sense of a realist construal of quantum mechanics which associates properties (observables) independently with each of two separated systems. If OLOC is violated, the observables associated with locally maximal operators are not themselves 'local' at all. Hence the question whether such observables can have their values changed by an environmental change, in the way denied by imposing ELOC, is not really a locality issue. The violation of ELOC becomes a locality issue only if OLOC obtains. In such a case, violation of ELOC shows that the value of an observable that may properly be said to pertain to one of two separated systems can have its value changed by altering the setting of an apparatus interacting with the other system. The whole discussion is similar to the distinctions drawn in the case of stochastic hidden-variable theories in Section 4.4, when violation of the completeness condition (4.32) is interpreted in terms of holism or nonseparability. The reason why ELOC and OLOC, although conceptually quite distinct, have apparently been conflated in the literature is that violation of either principle, when expressed in terms of measurement results, demonstrates a dependence of the outcome recorded by the apparatus connected to one system on the setting of the apparatus connected to the other (remote) system.

If we decide to retain OLOC, then our argument provides a demonstration that ELOC is violated which has a quite different character from that involved in discussions of the Bell inequality. Here, the very considerable literature has concentrated on the assumptions implicit in the derivation of various forms of the Bell inequality. Our approach in this chapter has made minimal and transparent use of probability theory, and is certainly free of any joint distribution assumption for observables associated with non-commuting operators. It is probably as close as we can get to a purely algebraic proof of nonlocality in realistic-construals of QM. Again, it should be noted that the use of counterfactuals in deriving the contradiction means that this proof of nonlocality only goes through under the assumption of a *deterministic* dependence of the values of observables on the environmental context. This is a similar limitation to that indicated in the discussion of the Eberhard-type proof of the Bell inequality given in Sections 4.1 and 4.2.

Notes and References

The results in this chapter are based on Heywood and Redhead (1983). As mentioned, the suggestion of investigating the extension of

Maczynski's theorem to locally maximal observables was made in Bub (1976). The proof that such an extension is possible can be found in Demopoulos (1980), section V, addenda. See also the discussion of flaws in the original version of Demopoulos's argument (with the opposite conclusion!) given in Humphreys (1980) and Bub (1980). The distinction between (ontological) separability and locality is given further emphasis in Howard (1985).

7

Realism and Quantum Logic

7.1. The Revisability of Logic

There is a large body of literature in the philosophy of QM which claims that the real conceptual revolution that should be recognized as engendered by consideration of the 'paradoxes' of QM is a revision of logic. Putnam, for example, sees logic as bearing the same relation to QM as geometry does to general relativity (GR). Expressed schematically

$$\frac{\text{geometry}}{\text{GR}} = \frac{\text{logic}}{\text{QM}}$$

Just as a physical theory, GR, led to a new conception of geometry, so QM should be seen as leading to a new conception of logic.

The basic idea is this. If L denotes the old classical logic, then we seem driven to new 'paradoxical' physical conceptions of potentiality, nonlocality, and so on. Let us denote this paradoxical 'new' physics by P'. Then the total theory we are dealing with is $L + P'$. But could we get back to our 'old', nonparadoxical physics of classical realism—call it P—by changing L to a 'new' logic L'? Schematically we would have the equation

$$L + P' = L' + P$$

In the spirit of the holistic-conventionalist approach to the philosophy of science, captured in an extreme version of the Duhem–Quine thesis, we have a free choice in the matter. Either we stick with the classical logic L and put up with paradoxical physics P', or we get rid of the paradoxes at the price of adopting a new logic L'.

Clearly, in this programme we are thinking of logic as capturing a special sort of proposition—the logical truths—and then we are going to blur the distinction between logical truths and other true propositions about the world. We are not thinking of logic as primarily concerned with the notion of valid consequence, of transmitting truth from the premisses to the conclusion in a valid argument. Of course, in classical logic the notion of valid consequence is reducible to the notion of logical truth via the semantic version of the so-called Deduction Metatheorem. (This will be explained more fully in Section 7.4.)

Let us review briefly some of the main approaches to understanding the notion of logical truth. Consider the view that logical truths are true in virtue of the meanings of the logical particles such as quantifiers and connectives. In the crude Platonist approach, these meanings attach to external absolute ideas, and are known to us by rational insight or intuition. There is no room here for revision of logic—a 'revised' logic would just be a 'wrong' logic. Opposed to this view is the empiricist approach to logic. The logical truths are just very general—indeed the most general—laws of physics. Their generality lies in the fact that they apply to any physical object or system whatever. We can say again, if we like, that the logical truths are true in virtue of the meanings of the logical particles; but these meanings are now derived from more fundamental correlatives in the physical world.

It follows from this view that, if we change our physics, we may be *forced* to change our logic, a much stronger claim than that of the conventionalist. But tying logic to physics in this way may make it difficult to understand how logic operates in other areas of discourse, such as mathematics, unless one holds to a totally empiricist position in epistemology, and regards all factual propositions as ultimately reducible to physics.

We have exposed here a tension between the revisability of logic and its universality. If we want to say that logic is revisable, then we may want to give up its absolute or universal status and retreat to a relativist position. There may be different logics appropriate to different areas of discourse. The question of which is the unique 'right' logic does not arise. We are only interested in the question: is it right for the purpose in hand? This might be termed an instrumentalist view of logic. Two ways of approaching alternative logics can now be entertained:

1. We might investigate alternative logics in a purely formal way, devising instruments for possible but unspecified application.
2. We might look at each particular application and try to sort out the appropriate logic for that particular purpose.

In the technical development of logic, these two approaches may be characterized as the *syntactic* and *semantic* approaches respectively.

7.2. Classical Propositional Logic

To fix our ideas, let us consider the classical propositional calculus (CPC). We begin with some syntactic notions.

1. A denumerably infinite set of variables $P = \{p_n | n \in \omega\}$ where ω is the first transfinite ordinal.
2. Connectives, 'not' symbolized $\sim$
   'or' symbolized $\vee$
   'and' symbolized $\wedge$
3. Punctuation brackets (,).

A *string* is any finite sequence of symbols. A *well-formed formula* (wff) is an element of a distinguished set of (informally 'meaningful') strings. We denote this set by $F = \{\text{wff}\}$.

From F we can now select a finite subset A which we shall treat as *axioms*, so $A = \{\text{axioms}\}$, and introduce *rules of inference* that enable us to generate further elements of F from A. We shall call the elements of F so generated, including trivially the elements of A, the *theorem-set* $\mathfrak{T}$ relative to these axioms and rules of inference. So we have a formal deductive system

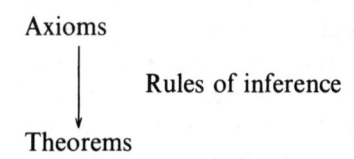

Axioms
↓ Rules of inference
Theorems

Now what is the object of doing all this in CPC? Well, the idea is to capture *all* the logical truths as elements of $\mathfrak{T}$. So from a small number of special logical truths—the axioms—we can systematically generate all the logical truths—the theorems. We want to show, in fact, the Completeness Metatheorem for CPC.

Completeness Metatheorem:
A wff is a logical truth iff it is a theorem.

(Note the somewhat idiosyncratic usage of completeness here to include the 'if' part of this result.)

To prove such a metatheorem, we need to have a clear definition of what we mean by a logical truth. This involves producing a *semantics* for CPC.

Classically, we proceed as follows. We introduce the notion of a *realization* as a map $f: F \to B_2$, where B_2 is the two-element Boolean algebra (we identify the unit in this algebra with Truth and the zero with Falsity). The map f is extended recursively from its restriction to the set P.

Thus
$$f(\phi \wedge \psi) = f(\phi) \wedge_2 f(\psi)$$
$$f(\phi \vee \psi) = f(\phi) \vee_2 f(\psi)$$
$$f(\sim \phi) = f(\phi)', \quad \forall \phi, \psi \in F$$

where $\wedge_2$, $\vee_2$ and $'$ are the meet, join, and complement operations in B_2.

Consider an arbitrary wff ϕ. We say that f satisfies ϕ iff $f(\phi) = 1$, in which case we write $f \models \phi$. ϕ is said to be *valid* or a *tautology*, the specialization of logical truth to the propositional calculus, if it is satisfied by all realizations. In this case we write $\models \phi$. The symbol for logical truth (validity) $\models$ should be compared with that for derivability (or provability) as a theorem in the syntactic axiomatization. Theoremhood is denoted by $\vdash \phi$.

The Completeness Metatheorem for CPC now says succinctly

$$\models \phi \text{ iff } \vdash \phi$$

Now let us consider how we might try to change our logic. First, we might try the purely formal syntactic approach, and just change some axioms and/or rules of inference so as to produce a different set of theorems. But we can raise the semantic objection to such a procedure, that the new set of theorems no longer capture the B_2-valid formulae. But we could try the opposite approach. Let us first of all change our notion of validity, and then set ourselves the formal task of producing an axiomatization whose theorems exactly coincide with all the valid formulae under the new notion of validity.

To proceed formally, let us replace B_2 by some other algebra A in the definition of realization, where we suppose that A is provided with suitable operations that can be made to correspond to the logical connectives. So we replace $f \in B_2^F$ by $f_A \in A^F$, and define

ϕ is A-valid iff

$$\forall f_A (f_A(\phi) \in D)$$

where D is some *designated set* of elements belonging to the algebra A. Of course it may happen that the A-valid formulae correspond exactly to the B_2-valid formulae. In this case we shall say that A is a *characteristic algebra* for CPC. With this terminology in mind, we shall now state and prove the following fundamental result.

Metatheorem (M):
Any arbitrary Boolean algebra is characteristic for CPC where the designated set is just the singleton whose sole member is the unit in the Boolean algebra.

Or, to put this another way: if B is an arbitrary Boolean algebra, any wff ϕ is B-valid iff it is B_2-valid.

Proof: We first note that if ϕ is B-valid then it must certainly be B_2-valid, since B always has B_2 as a sub-algebra. To show the converse, assume ϕ is B_2-valid but not B-valid, so under some B-realization F, ϕ is mapped onto an element $\neq 1$, i.e. $F(\phi) \neq 1$. Then $\sim \phi$ is mapped onto an element $\neq 0$, i.e. $F(\sim \phi) \neq 0$.

But by the Ultrafilter Theorem (see Mathematical Appendix, p. 178) it follows that $F(\sim \phi)$ is contained in some untrafilter.

Now we can use the Homomorphism Theorem (see p. 178) to show that there exists a homomorphism $H: B \to B_2$ which maps $F(\sim \phi)$ onto 1, i.e. $H(F(\sim \phi)) = 1$. But this implies $H(F(\phi)) = 0$. In other words, $f = H \circ F$ provides a B_2-realization for which $f(\phi) = 0$, so ϕ is not B_2-valid, contrary to hypothesis. So by *reductio* ϕ must be B-valid. Q.E.D.

So far we have introduced B-realizations or, more generally, A-realizations as though they were semantically on a par with B_2-realizations. In fact this is not the case. B_2-realizations are truth valuations, and are none other than the map Val introduced in Section 5.2 (p. 131). The two truth values encoded in B_2 are Fregean *references* which Val attaches to the wff s. But, as we shall see immediately, B- or A-realizations, if they are anything more than technical mathematical exercises, correspond to attaching Fregean *senses* to the wffs. They provide an intensional rather than an extensional semantics.

To illustrate what is going on, let us now show how B-realizations arise in the context of classical physics.

7.3. The Logic of Classical Physics

Classical physics represents the state of a physical system by a location in the relevant phase space. The observables or physical magnitudes are real-valued functions on the phase space. So, knowing the state, we can answer any question about the corresponding value of any physical magnitude. Classical physics is concerned with specifying the appropriate phase space for describing the state and, furthermore, specifying rules (equations of motion) which tell us how the state changes with time.

If we want to describe the global state of the whole universe, we can do this by constructing the Cartesian product of the phase spaces Ω_r for individual systems (fields or particles).

Thus
$$\Omega = \coprod_{r \in I} \Omega_r$$

where I is the set of all field and particles.

We introduce the notion of an *elementary proposition* as one which specifies the representative point of the universe as lying within a certain definite subset of Ω. Let us use small letters $p, q \ldots$ to represent elementary propositions, and capital letters $P, Q \ldots$ to represent the associated subsets of Ω. So the elementary proposition p says that the representative point of the universe is contained in the subset P of Ω, while the elementary proposition q says that the representative point of the universe is contained in the subset Q of Ω, and so on.

Let us now consider what meaning to attach to compound propositions. 'p or q' means that the representative point of the universe lies in the subset P or the subset Q, i.e. that it lies in the union $P \cup Q$ of the subsets P and Q. Similarly, 'p and q' means that the representative point lies in the intersection $P \cap Q$. Finally 'not p' means that the representative point lies outside of P, i.e. lies in $\mathscr{C}P$, the complement of P in Ω.

So we can set up equivalences between the logical connectives $\vee, \wedge$, and $\sim$, and the set-theoretic operators, $\cup, \cap$, and $\mathscr{C}$.

$$\vee \text{ corresponds to } \cup$$
$$\wedge \text{ corresponds to } \cap$$
$$\sim \text{ corresponds to } \mathscr{C}$$

Indeed, we can regard the meanings of $\vee, \wedge$, and $\sim$, in so far as these connectives are employed in classical physics, as derived from the antecedently understood Boolean operations on sets.

Normally, in presentations of set theory, one thinks of the Boolean operations as being defined in terms of the logical connectives. Thus

$$P \cup Q \underset{Df}{=} \{x | (x \in P) \vee (x \in Q)\}$$

$$P \cap Q \underset{Df}{=} \{x | (x \in P) \wedge (x \in Q)\}$$

$$\mathscr{C}P \underset{Df}{=} \{x | \sim (x \in P)\}$$

But now, in the spirit of an empiricist approach to logic, we are effectively reversing these definitions

$$(x \in P) \vee (x \in Q) \underset{Df}{=} x \in (P \cup Q)$$

$$(x \in P) \wedge (x \in Q) \underset{Df}{=} x \in (P \cap Q)$$

$$\sim (x \in P) \underset{Df}{=} x \in \mathscr{C}P.$$

We have seen how compound propositions constructed from elementary propositions yield new elementary propositions. What now corresponds to a logical truth or tautology? This is a proposition which says that the representative point lies somewhere in Ω, but gives no more precise delimitation. It is certainly a true statement about the physical world, but in a sense 'uninformative'. We do not have to look at the world to know that it is in *some* state.

Let us relate these remarks to the notion of a B-realization. We take for B simply the power set Boolean algebra associated with Ω. The unit in this algebra is just Ω itself. So the tautologies are just the wffs which always map onto Ω, i.e. the B-valid formulae. The first thing to notice is that, in view of our metatheorem (M), the class of tautologies picked out in this way is exactly the same as the class of tautologies picked out by means of truth valuations, i.e. B_2-realizations. In a sense, this fact explains why CPC is the appropriate logic for classical physics. Notice also that, having given a meaning, or sense, to the elementary propositions by means of the B-realization, we can go on to provide a truth valuation by mapping B homomorphically onto B_2. The construction of these homomorphisms is discussed in the Mathematical Appendix (pp. 177 ff.).

There is another point of considerable importance. We could just as well use Ω_r in place of Ω to spell out the tautologies. Remember that B-validity is equivalent to B_2-validity for *any* B, so the power set of Ω_r will do just as well as the power set of Ω. In other words, the same logic, viz. CPC, applies whether we are dealing with the whole universe or just a part of it.

Finally, we do not have to take for B the full power set of the phase space. To be technical for a moment, we could use the σ-algebra of Borel subsets, for example. In this case it should be noted that the Stone space of such a σ-algebra is not identical with the original phase space, a claim sometimes made in the literature, due to the presence of nonprincipal ultrafilters in the algebra (see Mathematical Appendix pp. 177 ff.).

7.1. Quantum Propositional Logic

The basic idea of quantum logic is to replace the Boolean lattice B appropriate to the phase space of classical physics by the projection lattice $\mathscr{L}$ of Hilbert space (cf. the discussion given in Section 1.4). The join and meet in this lattice correspond to the operations of 'linear span' and 'set-theoretical intersection', when we are thinking in terms of the ranges of the projection operators, viz. the subspaces of Hilbert space. The projection lattice is also equipped with an orthocomplement corresponding to the orthogonal subspace.

We now make the following equivalences between the logical connectives $\vee$, $\wedge$, and $\sim$ and the lattice operations of join, meet, and orthocomplement

$\vee$ corresponds to linear span $\oplus$

$\wedge$ corresponds to set-theoretic intersection $\cap$

$\sim$ corresponds to orthocomplement $\perp$

The quantum-logical tautologies are identified with the $\mathscr{L}$-valid wff, as compared with the B-valid wff of CPC.

But now we must be very careful. What are we going to mean by an *elementary proposition* in quantum logic (QL)? The first suggestion we shall make is that it is a proposition, $e(U)$, say, which asserts that the state vector of the system under discussion lies in the subspace U of the Hilbert space H. Let $u, v \ldots$ represent elementary propositions, and $U, V \ldots$ the corresponding subspaces of H. Then '$u \vee v$' means 'the state vector of the system lies in the subspace $U \oplus V$'. '$u \wedge v$' means 'the state vector of the system lies in the subspace $U \cap V$'. '$\sim u$' means 'the state vector of the system lies in the orthogonal subspace $U^\perp$'. A tautology is the true but uninformative proposition which asserts that the state vector of the system lies somewhere in H.

Having mapped the elementary propositions onto subspaces of H, we can now provide a truth valuation for our quantum propositional calculus by mapping subspaces onto truth values, elements of B_2, subject to the following admissibility criterion

A_1:

Val: $\{u\} \to B_2$ is an admissible valuation iff there is a one-dimensional subspace N such that for every subspace M, $\mathrm{Val}(m) = 1$ iff $N \leqslant M$.

Here $\leqslant$ is used to specify the subspace relation. Under the admissibility criterion A_1, the quantum-logical tautologies all get mapped onto the unit in B_2, as we should expect, and the state of the system is required to be a unit vector contained in N.

An important point should be noticed. We have been developing the idea of QL with reference to a particular Hilbert space describing some specified QM system. In general, the logic will depend on which Hilbert space we choose. The situation is unlike CPC, where B-validity for any B gives the same set of tautologies. We can define *minimal quantum logic* (MQL) in terms of $\mathscr{L}$-validity for the projection lattices of all possible Hilbert spaces (of different finite dimensions and infinite dimension). Notice, furthermore, that A_1-admissible truth valuations do not in general involve homomorphisms from $\mathscr{L}$ onto B_2. Indeed, for Hilbert spaces of dimensions $\geqslant 3$, such homomorphisms simply do not exist. This amounts to saying that the QL connectives are not truth functional, i.e. the truth value of a compound proposition is not determined by the truth values of the constituents. For example, if U and V are one-dimensional subspaces of $\mathbb{H}$, and suppose u and v, the associated elementary propositions, are both false, then $u \vee v$ is true if the QM state vector lies in $U \oplus V$, but false if it does not.

The most obvious feature that distinguishes such quantum logic from classical logic is the failure, in general, of the distributive rules such as

$$u \wedge (v \vee r) \equiv (u \wedge v) \vee (u \wedge r)$$

The symbol of logical equivalence ($\equiv$) means that, under any truth valuation, the two propositions get the same truth value. This clearly breaks down in QL owing to the fact that $\mathscr{L}$ is not a distributive lattice, as we demonstrated in Section 1.4.

In Section 7.3 we noted that in classical physics observables are real-valued functions on the phase space. Formally, we represent an observable Q as a map $Q: \Omega \to \mathbb{R}$. Consider now a *new* sort of elementary proposition: Q has a value lying in the set Δ. Adapting the notation of Section 5.2, we represent this statement as $(\Delta)_Q$. But we now have a simple logical equivalence between $(\Delta)_Q$ and the elementary proposition $e(P)$, which asserts that the representative point lies in the subset P of Ω

$$(\Delta)_Q \text{ iff } e(Q^{-1}(\Delta))$$

So, knowing the truth values for elementary propositions about the state of the system gives us immediate information about the truth values for the new sort of elementary proposition concerning the values of observables.

How does this work out in QL? In the case where $|\phi\rangle \in \text{ran}(P_Q(\Delta))$ we have that

$$\text{Prob}(\Delta)_Q^{|\phi\rangle} = 1$$

This is at any rate consistent with simply asserting the truth of the proposition $(\Delta)_Q$, interpreted realistically as asserting that the possessed value of Q lies in the set Δ. If we make this move, then, schematically, we can say the following:

$$e(U), \text{ for some } U \leqslant \text{ran}(P_Q(\Delta))$$
$$\to (\Delta)_Q$$

(Note that $\to$ is here being used as a metalinguistic symbol of material implication.)

This gives a sufficient condition for asserting the truth of $(\Delta)_Q$, but says nothing about necessary conditions, i.e. about sufficient conditions for asserting the falsity of $(\Delta)_Q$. We shall return to this matter in the next section. But first let us see how the QL connectives operating on the elementary propositions such as $e(U)$ translate in terms of propositions about the values of observables, such as $(\Delta)_Q$, in a situation where the translation can be made.

Consider some operator $\hat{Q}$ with discrete nondegenerate spectrum $\{q_1, q_2 \ldots\}$. Denote the subspaces generated by the eigenvectors $|q_1\rangle, |q_2\rangle \ldots$ by $Q_1, Q_2 \ldots$

Then $\qquad e(Q_1) \to (\{q_1\})_Q$
and $\qquad e(Q_2) \to (\{q_2\})_Q \qquad$ etc.

In words:

If the QM state lies in Q_1, Q has the value q_1.
If the QM state lies in Q_2, Q has the value q_2.
etc.

Then $e(Q_1) \lor e(Q_2) \equiv e(Q_1 \oplus Q_2)$
$$\to (\{q_1, q_2\})_Q$$
$$(\equiv (\{q_1\} \cup \{q_2\})_Q)$$

What this result says is that, if we assert the QL disjunction of $e(Q_1)$ and $e(Q_2)$, then we are licensed to assert that Q has the value q_1 *or* the value q_2, where 'or' is now being used in the *classical* sense.

In terms of measurement results, the QL disjunction says, that the result of a measurement is always either q_1 or q_2. This should be compared with the classical disjunction, which would assert that the result of measurement is either always q_1 or always q_2.

The contrast is even clearer for negation

$$\sim e(Q_1) \equiv e(Q_1^\perp)$$
$$\to (\mathscr{C}\{q_1\})_Q$$

where $\mathscr{C}\{q_1\}$ denotes the set-theoretic complement of $\{q_1\}$ in $\{q_1, q_2 \ldots\}$, i.e. $\{q_2, q_3 \ldots\}$.

In terms of measurement results the quantum negation implies that the result of a measurement of Q is never q_1, whereas classical negation would imply that the result of measurement was not always q_1.

In the account we have been giving, it is clear that QL is not a rival to classical logic, but an alternative for expressing different sorts of compound propositions. For example, disjunction involves a lack of specificity of the location of the QM state vector in Hilbert space, but of a different kind from that afforded by the classical disjunction. In fact, the meaning of the QL connectives which we 'read off' the projection lattice can always be 'translated' in terms of the set-theoretic structure of Hilbert space, and set-theoretic structures, as we have seen, involve classical logic.

We close this section with two remarks:

1. In QL the de Morgan laws are valid

$$u \wedge v \equiv \sim ((\sim u) \vee (\sim u))$$
and
$$u \vee v \equiv \sim ((\sim u) \wedge (\sim v))$$

This follows from the fact that for any subspaces U and V of a Hilbert space

$$(U^\perp \oplus V^\perp)^\perp = U \cap V$$
and
$$(U^\perp \cap V^\perp)^\perp = U \oplus V$$

This means that, in QL, we can define $\wedge$ in terms of $\vee$ and $\sim$, or $\vee$ in terms of $\wedge$ and $\sim$, just as in classical logic.

2. We have so far talked exclusively about the notion of validity in relation to logical truth, rather than logical entailment. But we can define: u logically entails v, symbolised $u \models v$, iff for all realizations $f_\mathscr{L}$, $\quad f_\mathscr{L}(u) \leqslant f_\mathscr{L}(v)$.

In terms of admissible valuations, this means that any valuation which makes u true will make v true. Now, in CPC, the *relation* of logical entailment can be reduced to the notion of logical truth via the operation of material implication.

Formally $\qquad\qquad p \models q$ iff $\models p \to q$

and $\qquad\qquad\qquad p \to q \equiv (\sim p) \vee q.$

There is nothing corresponding to this so-called Deduction Metatheorem in QL, since there is no connective that plays exactly the role of material implication.

7.5. Putnam States and Realism

We now return to considering the elementary proposition $(\Delta)_Q$, which asserts that the observable Q possesses a value which lies in the subset Δ of possible values for Q. In a realistic interpretation of QM, one wants all such propositions to have well-defined truth values. The following scheme has been proposed (effectively by Putnam). First of all map $(\Delta)_Q$ onto $P_Q(\Delta)$, and then associate with $P_Q(\Delta)$ the proposition

∗ 'The state $|\phi\rangle$ of the system lies in the range of $P_Q(\Delta)$'

We have already argued that, if the QM state of the system were to be such a $|\phi\rangle$, then we could assert the truth of $(\Delta)_Q$. But suppose the QM state does not lie in the range of $P_Q(\Delta)$. We may still want to assert the truth of $(\Delta)_Q$. But what then is the significance of the state $|\phi\rangle$ in the proposition∗? We shall refer to it as the 'real' state or Putnam state associated with the system. It must be sharply distinguished from the QM state. Corresponding to the elementary proposition $e(U)$, which asserted that the QM state lies in the subspace U of $\mathbb{H}$, we now introduce still another sort of elementary proposition, symbolized as $e'(U)$, which asserts that the Putnam state lies in the subspace U of $\mathbb{H}$. In particular, the proposition $(\Delta)_Q$ is translated as $e'(\text{ran}(P_Q(\Delta)))$.

Quantum logic for the propositions $(\Delta)_Q$ is now introduced via the QL connectives for the e'-propositions specifying the location of the Putnam state, in exact correspondence with the way we treated e-propositions specifying the location of QM states, in the previous section. Again, letting $u, v, \ldots$ now represent the elementary propositions $e'(U), e'(V)$, we seek a truth valuation which will reflect our realist intentions. It would appear that we require the following admissibility criterion for such a valuation

A_2:

Val: $\{u\} \to B_2$ is an admissible valuation iff the following conditions are satisfied
1. Val $(u) = 1$ iff val $(\sim u) = 0$
2. If Val $(n) = 1$, and $N \leq M$, then Val $(m) = 1$
3. In any orthonormal basis $\{|q_i\rangle\}$ of ℍ, where $\{q_i\}$ are the eigenvalues of some maximal observable Q, Val $(q_j) = 1$ for some j and hence from (1) and (2) Val $(q_i) = 0$, $\forall i \neq j$.

Here q_j is a convenient shorthand for the proposition $e'(Q_j)$, where Q_j is, as before, the one-dimensional subspace generated by $|q_j\rangle$. It also represents the proposition $(q_j)_Q$.

The first condition in A_2 is required in order to guarantee that, for any observable Q (maximal or not), $(\Delta)_Q$ is true iff $(\mathscr{C}\Delta)_Q$ is false, where $\mathscr{C}\Delta$ is the complement of Δ in the set of possible values for Q.

The second condition in A_2 ensures that

$$(\Delta)_Q \to (\Delta')_Q \text{ if } \Delta \subseteq \Delta'.$$

The third condition is the crucial one which ensures that every maximal observable Q has a unique definite value. It will also follow that nonmaximal observables have unique definite values, since every subspace associated with degenerate eigenvectors is mapped onto a definite truth value. If it is 1, all other eigenspaces get mapped onto 0, by applying condition (1), and one eigenspace certainly is mapped onto 1, using (3) and (2). But there are immediate problems with A_2:

1. The condition (3) shows that the system has many different Putnam states simultaneously—indeed, a different state for every distinct orthonormal basis! This shows that the term 'state' is being used in a very Pickwickian sense; and the justification for QL applied to propositions about the location of Putnam states in Hilbert space, in analogy with the location of QM states, is quite inappropriate. In fact we might just as well drop all talk of Putnam states, and simply impose QL directly on the $(\Delta)_Q$ type of proposition. There is really no intuitive justification for doing this, but let us see where it will take us.

2. However, a more serious difficulty is that there are no A_2-admissible valuations for Hilbert spaces of dimension ≥ 3. This follows directly from the Two-Colour Theorem proved in Section 5.1, where the colour map is replaced by Val. Does this mean that QL does not, after all, help with the realism issue? Putnam has presented the

following argument. In QL, realism means that every observable has a value, but there is no value which it has!

Consider a maximal observable Q in a Hilbert space of dimension N. Then Putnam identifies the statement 'Q has a value' with the disjunction $q_1 \vee q_2 \vee \ldots q_N$. This is not only true but is a QL tautology, since it corresponds to the whole Hilbert space ($Q_1 \oplus Q_2 \oplus \ldots Q_N = \mathbb{H}$). However, says Putnam, this does not mean that there is some specifiable j for which q_j is true. Now, in classical logic, the disjunction carries an existential commitment. We can write

$$(\exists i) q_i \equiv q_1 \vee q_2 \vee \ldots q_N$$

Putnam is effectively retaining this classical result, but using it to *define* what he means by $(\exists i) q_i$. In QL, then, $(\exists i) q_i$ does *not* mean that an i can be specified for which q_i is true. What does it mean? Well just $q_1 \vee q_2 \vee \ldots q_N$. But then, we might ask, what has the truth of $(\exists i) q_i$ got to do with a realism of possessed values?

Consider now some other maximal observable R with eigenvectors $\{|r_i\rangle\}$ distinct from $\{|q_i\rangle\}$. So $\hat{Q}$ and $\hat{R}$ do not commute. Nevertheless

$$(\exists j) r_j \underset{Df}{=} r_1 \vee r_2 \vee \ldots r_N$$

is also tautologically true.

Putnam asks us to consider the two statements

$$S_1 : (\exists i) q_i \wedge (\exists j) r_j$$
and
$$S_2 : (\exists i)(\exists j)(q_i \wedge r_j)$$

In classical logic S_1 and S_2 are equivalent as a result of the Distributive Law. This can be seen at once by writing S_1 and S_2 in the expanded forms

$$S_1 : (q_1 \vee q_2 \ldots \vee q_N) \wedge (r_1 \vee r_2 \ldots \vee r_N)$$
$$S_2 : (q_1 \wedge r_1) \vee (q_1 \wedge r_2) \ldots \vee (q_1 \wedge r_N)$$
$$\vee (q_2 \wedge r_1) \vee (q_2 \wedge r_2) \ldots \vee (q_2 \wedge r_N)$$
$$\vdots$$
$$\vee (q_N \wedge r_1) \vee (q_N \wedge r_2) \ldots \vee (q_N \wedge r_N)$$

But, in QL, S_1 and S_2 are not logically equivalent. S_1 is a tautology while S_2 is a logical contradiction!

Now, Putnam remarks, S_1 reflects realism—that is why it is true, indeed tautologically so—while the fact that S_2 is a logical contradiction expresses complementarity: it is logically contradictory to say that the system possesses simultaneous values for Q and R.

But is this sort of talk genuinely helpful in understanding QM? It may be argued that, far from helping to resolve the mysteries in QM, it merely substitutes one mystery for another, viz. how to make sense of the sort of slogans we have been repeating. Any attempt to produce a proper realist semantics for QL, via A_2-admissible valuations, for example, seems to run foul of the Kochen–Specker paradox. But if we don't do that, can we really be said to be retaining realism?

Notes and References

A general introductory reference on the philosophy of logic is Haack (1978). The notion of a characteristic algebra (or matrix) was originally introduced by Lindenbaum and Tarski in order to capture algebraically the notion of theoremhood. We have adapted the terminology in a semantic rather than a syntactic context. For technical details on classical logic, the best reference is Bell and Machover (1977).

We have been concerned in this chapter with a very specific type of bivalent nondistributive quantum logic. The original references that may be consulted include Birkhoff and von Neumann (1936), Putnam (1969), and Putnam (1974). We have been concerned with Putnam's views as he expressed them in these references. His own views have since been modified. See, for example, Friedman and Putnam (1978) and Putnam (1981). On the question of admissible truth valuations we have followed Friedman and Glymour (1972). For the general setting of different varieties of quantum logic, see in particular chapter 8 of Jammer (1974), van Fraassen (1974), Beltrametti and van Fraassen (1981), and Holdsworth and Hooker (1983).

A useful philosophical critique of Putnam's version of QL is provided by Dummett (1976). The notion of minimal QL, referred to in the text, is expounded in Goldblatt (1974). Another perceptive reference on QL is Bell and Hallett (1982). For a clear discussion of how far one can get in defining a surrogate material conditional in QL, see Hardegree (1974).

8
Envoi

IN this book we have been mainly concerned with the difficulties encountered by a simple-minded realism of possessed values. The reaction to these difficulties on the part of many philosophers is to say we must not be so simple-minded. The quantum logician, for example, refuses to allow the problem posed for a realist semantics by the Kochen–Specker paradox by simply denying that realism commits him to a specifiable truth valuation for all elementary propositions ascribing particular values to observables.

Again, the complementarity interpretation we labelled view C in Chapter 2 is in effect denying separability, in the sense that systems possess properties independently of a holistic context. Denial of separability in this ontological sense would also block the derivation of the Bell inequality given in Chapter 4, and gives one way of avoiding the contradiction derived in Chapter 6. Separability was there referred to as the OLOC principle.

If we want to *be* simple-minded and keep to a separable realism of possessed values, then the arguments given in Chapter 4 tell us, I would submit rather convincingly, that action-at-a-distance (what we called a violation of the ELOC principle in Chapter 6) must be admitted. This might be thought to raise severe problems for special relativity. But the fact that no controllable information can be transmitted by employing such nonlocality is a very important point to bear in mind in this connection. Alternatively, on view B of Chapter 2, the original EPR argument, as presented in Chapter 3, again leads to nonlocality (violation of LOC_1). But this form of nonlocality, actualizing possibilities at a distance, may be thought sufficiently far removed from the classical stamping ground of special relativity to allow what Shimony has referred to as 'peaceful coexistence'.

Another important issue discussed in Chapter 4 was the rôle of determinism in the derivation of the Bell inequality. Our conclusion here was that determinism *was* needed to justify certain counterfactual moves made in the derivation. But if we give up determinism,

Envoi 169

then the discussion of Section 4.4 on stochastic hidden-variable theories shows that requirements of locality and completeness (again effectively equivalent to separability) lead to the Bell inequality. In addition, factorizable stochastic theories cannot anyway accommodate the cases of strict correlation or anti-correlation. In Section 4.4 we discussed how nonseparability equated with 'passion-at-a-distance'. Again this might be regarded as compatible with the constraints of special relativity, interpreted as prohibiting the operation of *causal* processes outside the light-cone, i.e. at space-like separation.

So there it is—some sort of action-at-a-distance or (conceptually distinct) nonseparability seems built into any reasonable attempt to understand the quantum view of reality. As Popper has remarked, our theories are 'nets designed by us to catch the world'. We had better face up to the fact that quantum mechanics has landed some pretty queer fish.

Notes and References

For recent philosophical discussions of realism see Leplin (1983). Peaceful coexistence between nonlocality and special relativity was mooted by Shimony (1978). For further discussion of the relevance of nonlocality to the foundations of special relativity, see in particular Redhead (1983) and (1986b). The quotation from Popper can be found in Popper (1982a), p. 42. *Cp.* also the motto from Novalis to *Logik der Forschung* (Popper (1934)).

Stapp has continued to try and develop a proof of the Bell inequality that does not assume determinism or a commitment to joint probability distributions for incompatible observables. The most recent version is Stapp (1988). For a critique see Clifton et al. (1988).

Further discussion of the no-signalling theorem is provided in Clifton and Redhead (1988), where a proposed signalling scheme involving CP violation is refuted.

Mathematical Appendix

I. Naive Set Theory

A *set*, S, is a collection of objects, termed the *members* of S, and is determined by its members in the sense that sets with identical members are regarded as identical. If x is a member of S, we write $x \in S$. A set with a finite number of members $x_1, x_2 \ldots x_n$ will be written as $\{x_1, x_2 \ldots x_n\}$. More generally, a set will be determined by some property P shared by its members. The set of all objects having the property P will be denoted by $\{x | P(x)\}$. Generating sets by this device in general leads to celebrated contradictions. We shall assume that all the sets considered are not sufficiently 'large' to be contradictory.

By the *union* of two sets A and B, we understand the set

$$A \cup B \underset{Df}{=} \{x | x \in A \text{ or } x \in B\}$$

where the inclusive sense of 'or' is understood. By the *intersection* of two sets A and B, we understand the set

$$A \cap B \underset{Df}{=} \{x | x \in A \text{ and } x \in B\}$$

By the *complement* of a set A, we understand the set

$$\mathscr{C}A \underset{Df}{=} \{x | x \notin A\}$$

where $\notin$ signifies 'is not a member of'. We also denote by $\varnothing$ the *empty set* which possesses no members.

A set A is said to be a *subset* of B if $\forall x$

$$x \in A \to x \in B$$

In this case we write $A \subseteq B$. $\subseteq$ will be referred to as the relation of set-theoretic inclusion.

If A is not identical with B it is called a *proper* subset of B, denoted by $A \subset B$.

It is conventional to regard the empty set $\varnothing$ as a proper subset of every other set.

The set of all subsets of any set A is called the *power set* of A.

We represent an ordered n-tuple as

$$(x_1, x_2, \ldots, x_n)$$

The *Cartesian product* of n sets $A_1 A_2 \ldots A_n$ is then the set

$$A_1 \times A_2 \times \ldots \times A_n = \{(x_1, x_2, \ldots, x_n) | x_1 \in A_1 \text{ and } x_2 \in A_2 \ldots \text{and } x_n \in A_n\}$$

Mathematical Appendix

An n-ary relation among the sets $A_1, A_2 \ldots A_n$ is just a subset of the Cartesian product $A_1 \times A_1 \ldots \times A_n$.

A special sort of binary relation between two sets is of particular importance. A *map* or *function* between two sets A and B written $f: A \to B$ is just a set of ordered pairs of the form (x, y) where $x \in A$ and $y \in B$. x ranges over the whole set A, and with each x is associated a unique y, the image of x under f which we write as $f(x)$. If each y is also associated with a unique x the function is said to be 1:1. Otherwise it is many-one.

The set A is called the *domain* of f, (dom (f)). The image of the domain is called the *range* of f, (ran (f)), where the image of a set of values in the domain is just the set of images defined in the point-wise fashion as above. ran (f) is in general a proper subset of B. If ran $f = B$, the function is said to be onto B rather than just into B.

Consider any set $Y \subseteq \text{ran}(f)$. We define the *inverse image* of Y under f as the set $\{x | x \in A \text{ and } f(x) \in Y\}$. We write this as $f^{-1}(Y)$. This notation does not mean that we can associate with every function f an *inverse function* f^{-1}. This is only true if f is 1:1. In that case, f^{-1} is defined by $f^{-1}(y) = x$, where $y \varepsilon$ ran (f), $x \in \text{dom}(f)$, and $y = f(x)$. Notice carefully that inverse images always exist, even when f is many-one so that no inverse function exists.

Consider any set $X \subseteq A$. Then we define the *restriction* of f to the set X, written $f|X$, as the function $f|X: X \to B$, such that $f|X(x) = f(x)$, $\forall x \in X$. We denote by B^A the set of all functions from A to B. The composition $g \circ f: A \to C$ of two functions is defined by $(g \circ f)(x) = g(f(x))$ where f maps $A \to B$ and g maps $B \to C$ and $x \in A$.

Using the notion of function, we can generalize the idea of a Cartesian product in the following way. Consider a family of sets $\{A_i : i \in I\}$ indexed by $i \in I$. We then define

$$\prod_i A_i = \{f: I \to \bigcup_i A_i | f(i) \in A_i\}$$

where $\bigcup_i A_i$ is the smallest set that contains as members all the members of all the sets A_i. $\prod_i A_i$ and $\bigcup_i A_i$ are then generalizations of Cartesian product and set-theoretic union for arbitrary collections of sets.

II. Finite-Dimensional Vector Spaces

A Vector Space is an ordered pair (V, K) where V is a set of elements, the *vectors*, and K is a *field* whose elements are termed *scalars*. On V there is defined an *internal binary operation* of *addition*, denoted by $+$, which is associative and commutative; and there exists a distinguished element of V, the *zero* vector, written 0, with the property

$$0 + x = x, \quad \forall x \in V$$

To every element of V there also corresponds an *inverse* in the sense

$$\forall x \in V, \exists y \text{ such that } x + y = 0$$

There is also defined an external binary operation of *scalar multiplication* with the properties

If $\qquad\qquad\qquad \lambda, \mu \in K; x, y \in V$

then $\qquad\qquad\qquad \lambda x \in V$

$$\lambda(x + y) = \lambda x + \lambda y$$
$$(\lambda + \mu)x = \lambda x + \mu x$$
$$(\lambda \mu)x = \lambda(\mu x)$$

A *linear operator* on a vector space is a *map* $Q: V \to V$ such that

$$Q(x_1 + x_2) = Q(x_1) + Q(x_2), \forall x_1, x_2 \in V$$

and $\qquad\qquad Q(\lambda x) = \lambda Q(x), \forall \lambda \in K, \forall x \in V.$

An *inner product* on a vector space is a scalar-valued function defined on the Cartesian product $V \times V$, i.e. a function $\phi: V \times V \to K$, where we write $\phi((x, y)) = \langle x|y \rangle; x, y \in V$. We restrict our attention henceforth to the case where K is the field of complex numbers. The inner product has the following properties

$$\langle x|y \rangle = \langle y|x \rangle^*$$

where * denotes complex conjugate.

$$\langle x|x \rangle \geq 0$$
$$\langle x|x \rangle = 0 \text{ if and only if } x = 0.$$
$$\langle x|\lambda_1 y_1 + \lambda_2 y_2 \rangle = \lambda_1 \langle x|y_1 \rangle + \lambda_2 \langle x|y_2 \rangle$$

and $\qquad \langle \lambda_1 x_1 + \lambda_2 x_2|y \rangle = \lambda_1^* \langle x_1|y \rangle + \lambda_2^* \langle x_2|y \rangle$

$\|x\| \overline{\underset{Df}{=}} (\langle x|x \rangle)^{\frac{1}{2}}$ is referred to as the *norm* of x.

A *linear functional* on a vector space is a map $\alpha: V \to K$ with the properties

$$\alpha(\lambda_1 x_1 + \lambda_2 x_2) = \lambda_1 \alpha(x_1) + \lambda_2 \alpha(x_2), \forall \lambda_1, \lambda_2 \in K, \forall x_1, x_2 \in V.$$

The set of linear functionals on a vector space is itself a vector space known as the *dual* space to V, written V'. If $\alpha \in V'$ and $x \in V$, a 1:1 correspondence ϕ between V' and V is defined by writing

$$\alpha(y) = \langle x|y \rangle, y \in V \text{ and then taking } x = \phi(\alpha).$$

It is a theorem that for any α there exists a unique x satisfying the defining relation.

A *complete orthonormal* set of *basis* vectors for V is a set $\{e_1 \ldots e_N\}$ such that any $x \in V$ can be written in the form

$$x = \sum_i^N \lambda_i e_i, \quad \lambda_i \in K$$

Mathematical Appendix

and the e_i satisfy the orthonormality condition

$$\langle e_i | e_j \rangle = \delta_{ij}$$

where δ_{ij} is the usual Kronecker delta symbol

$$\delta_{ij} = 1 \text{ for } i = j$$
$$= 0 \text{ for } i \neq j$$

The number N is the dimension of the vector space.

Note that in the text we follow the Dirac notation of enclosing vectors with the ket symbol $| \rangle$. $Q(x)$ is then written $Q|x\rangle$.

The *matrix representation* of a linear operator Q is the $N \times N$ matrix Q_{ij} defined by

$$Q_{ij} = \langle e_i | Q(e_j) \rangle$$

The *column matrix* representing the vector x is just

$$\begin{pmatrix} \lambda_1 \\ \lambda_2 \\ \vdots \\ \lambda_N \end{pmatrix}$$

We have then the obvious relations

$$Q e_j = \sum_{i=1}^{N} Q_{ij} e_i$$

and

$$Q x = \sum_{i=1}^{N} \mu_i e_i$$

where

$$\mu_i = \sum_{j=1}^{N} Q_{ij} \lambda_j$$

or in matrix notation

$$\begin{pmatrix} \mu_1 \\ \vdots \\ \mu_N \end{pmatrix} = \begin{pmatrix} Q_{11} & \cdots & Q_{1N} \\ \vdots & & \vdots \\ Q_{N1} & \cdots & Q_{NN} \end{pmatrix} \begin{pmatrix} \lambda_1 \\ \vdots \\ \lambda_N \end{pmatrix}$$

The *adjoint* of a linear operator Q is written $Q^\dagger$ and defined by the relation

$$\langle x | Q^\dagger(y) \rangle = \langle y | Q(x) \rangle^*, \quad \forall x, y \in V$$

An operator is *self-adjoint* if $Q = Q^\dagger$. An operator is *unitary* if $Q^\dagger = Q^{-1}$ where Q^{-1} is the inverse map to Q. (Every operator has an adjoint, but of course it is not the case that every operator has an inverse). The *eigenvalues* of Q are the numbers q_α satisfying

$$Q(x) = q_\alpha x, \text{ for some } x \in V.$$

x is said to be an *eigenvector* belonging to the eigenvalue q_α. There are at most N distinct eigenvalues. If Q is self-adjoint, it is a theorem that all the q_α are real. In this case we can always choose a set $\{x_i\}$ of eigenvectors, $i = 1, 2 \ldots N$, which provide a complete orthonormal basis for V. We denote the eigenvalue associated with x_i by q_i. Some of the q_i may be equal. In this case, Q is said to be degenerate. If Q is nondegenerate, so that there are N distinct eigenvalues, it is often referred to as *maximal*. In Dirac notation we write x_i as $|q_i\rangle$.

A *projection operator* is a self-adjoint *idempotent* linear operator. If P is a projection operator, the idempotent property is expressed by

$$P^2 = P$$

The eigenvalues of P are easily seen to be 0 or 1. It is a theorem that the range of a projection operator is a *subspace* of the vector space, where a subspace U is a *subset* of V, which is itself closed under the operations of addition and scalar multiplication. If $P_{|q_i\rangle}$ is the projection operator whose range is the subspace spanned by $|q_i\rangle$ (a subspace is *spanned* by a set of vectors if every element of the subspace can be expressed as a linear combination of members of this set), then the *spectral theorem* reads

$$Q = \sum_{i=1}^{N} q_i P_{|q_i\rangle}$$

a result which is proved in the text.

The $P_{|q_i\rangle}$ satisfy the orthogonality relation

$$P_{|q_i\rangle} \cdot P_{|q_j\rangle} = \delta_{ij} P_{|q_i\rangle}$$

A function of a self-adjoint operator Q is defined by the relation

$$f(Q) \underset{Df}{=} \sum_{i=1}^{N} f(q_i) P_{|q_i\rangle}$$

It is a theorem that if two self-adjoint operators Q and R commute, then there exists a maximal self-adjoint operator S such that

$$Q = f(S)$$
$$R = g(S)$$

for appropriate functions f and g. On the other hand if $Q = f(S)$ and the map f is not 1:1, then there will always exist another maximal operator R such that $Q = g(R)$, and S and R do not commute. These results are proved in the text. If two operators commute, it is always possible to choose a basis relative to which they are simultaneously diagonal. If two operators fail to commute it does not follow that in every case they have no common eigenvectors, just that all their eigenvectors cannot be common to both operators.

The *tensor product of two vectors* v and w selected from vector spaces V and

Mathematical Appendix

W is written $v \otimes w$ and satisfies the relation

$$\left(\sum_i a_i v_i\right) \otimes \left(\sum_i h_i w_i\right) = \sum_{i,j} a_i h_j v_i \otimes w_j$$

$\forall v_i \in V, \ \forall w_i \in W$ and $\forall a_i, b_i \in K$

The *tensor product of two vector spaces* V and W, written $V \otimes W$, is the set of all linear combinations of the tensor products of vectors selected from the two spaces. (More formally $V \otimes W$ is the space of bilinear functionals defined over the Cartesian product of the dual spaces V' and W'.) $V \otimes W$ is itself a vector space with a basis $\{e_i \otimes e'_j\}$ where $\{e_i\}$ is a basis for V and $\{e'_j\}$ is a basis for W.

The *tensor product of two linear* operators A and B where A acts on V and B on W, is written $A \otimes B$. It is itself a linear operator on $V \otimes W$ and is defined by the relation

$$(A \otimes B)(v \otimes w) = A(v) \otimes B(w), \ \forall v \in V, \forall w \in W.$$

If A and C act on V, and B and D act on W, then we note the very useful relation:

$$(A \otimes B)(C \otimes D) = (AC) \otimes (BD)$$

which follows immediately from this definition.

As a special case of this result

$$(A \otimes I)(I \otimes B) = A \otimes B$$

where I denotes the *identity* operator, which maps each vector onto itself. Note also, for self-adjoint operators A and B,

$$f(A \otimes I) = f(A) \otimes I$$

and similarly

$$g(I \otimes B) = I \otimes g(B)$$

for any functions f and g.

A Note on Infinite-Dimensional Vector Spaces

For finite-dimensional vector spaces, all concrete realizations are isomorphic in respect of their vector space structure, so there is effectively only one abstract vector space of given dimension N. As N becomes infinite, the situation is more complicated. In order to isolate a categorical structure (i.e. all of whose realizations are isomorphic) it is necessary to impose additional axioms of a topological nature. In the case of the von Neumann formulation of QM, the structure singled out is called *Hilbert Space*.

Hilbert Space is an inner product vector space which satisfies conditions of *separability* (there is a denumerably infinite set of vectors which are *dense* in the whole space) and *completeness* (every Cauchy sequence of vectors in the space converges to a limit which is also a member of the space). Every finite-dimensional inner-product space is automatically a Hilbert space, since these additional axioms are now rather trivial theorems. But in the infinite-dimensional case, we cannot prove separability and completeness from the other axioms: we have to postulate them. Two famous concrete realizations of the infinite-dimensional Hilbert space are the space L_2 of square-integrable *functions* (integrable, that is, in the Lebesgue sense) and the space l_2 of square-summable *sequences*. The isomorphism between L_2 and l_2 in respect of their Hilbert space structure is the mathematical reflection of the equivalence between Schrödinger's wave mechanics and Heisenberg's matrix mechanics. It is known in the mathematical literature as the essential content of the *Riesz–Fischer theorem*.

III. Lattice Theory

A *partially ordered set* (poset) is a set S equipped with a binary relation $\leq$ that is

1. reflexive: $a \leq a$, $\forall a \in S$.
2. antisymmetric: $a \leq b$ and $b \leq a \to a = b$, $\forall a, b \in S$.
3. transitive: $a \leq b$ and $b \leq c \to a \leq c$, $\forall a, b, c \in S$.

We denote the *least upper bound* or *join* of two elements a and b by $a \vee b$, and the *greatest lower bound* or *meet* by $a \wedge b$, where the least upper bound means the 'lowest' element c in the partial order, such that $a \leq c$ and $b \leq c$; similarly, the greatest lower bound is the 'highest' element c, such that $c \leq a$ and $c \leq b$.

A *lattice* is a partially ordered set such that every pair of elements possess both a least upper bound and a greatest lower bound. A lattice possesses a zero and unit element, denoted by 0 and 1 respectively, if $0 \leq a$ and $a \leq 1$, $\forall a \in S$. A *complement* of an element a in a lattice is an element a' such $a \wedge a' = 0$ and $a \vee a' = 1$. A lattice is said to be *complemented* if every element possesses a complement (in general the complement is not unique). A lattice is *distributive* if $\forall a, b, c \in S$

$$a \wedge (b \vee c) = (a \wedge b) \vee (a \wedge c)$$
$$a \vee (b \wedge c) = (a \vee b) \wedge (a \vee c)$$

A complemented distributive lattice is known as a *Boolean* lattice, commonly referred to as a *Boolean algebra*. The best known example of a Boolean algebra is a power set algebra, i.e. the algebra obtained by taking the power set of a set, which is partially ordered by set-theoretic inclusion, the least upper bound of two elements in the algebra being the union of the subsets, the greatest lower bound the intersection of the subsets, and the complement the set-theoretic complement relative to the original set. As we shall see in a moment, all finite Boolean algebras are structurally equivalent (i.e. isomorphic) to power set algebras.

In a Boolean algebra each element has a unique complement, which satisfies the following additional conditions for being an *orthocomplement*.

1. $(a')' = a$
2. $a \leq b \leftrightarrow b' \leq a'$.

The projection lattice of a Hilbert space referred to in the text is ortho complemented but nondistributive.

Homomorphisms of Boolean Algebras

A homomorphism between two Boolean algebras B and B' is a map $h: B \to B'$, such that the Boolean operations are preserved

$$h(a \wedge b) = h(a) \wedge h(b)$$
$$h(a \vee b) = h(a) \vee h(b)$$
$$h(a') = h(a)'$$
$$h(0) = 0$$
$$h(1) = 1$$
$$a \leq b \leftrightarrow h(a) \leq h(b)$$

On the LHS of these results, the Boolean operations, zero and unit, refer to B, and on the RHS to B'. (Note that only the third property and one or the other of the first two are required to specify a homomorphism. The remaining properties can then be demonstrated.) We are particularly interested in *two-valued homomorphisms* of a Boolean algebra B which are homomorphisms from B onto the two-element Boolean algebra B_2, which consists just of a zero and a unit. B_2 is the only Boolean algebra in which the ordering is total rather than partial (a *total* ordering is one for which $a \leq b$ or $b \leq a$, $\forall a, b \in S$). We now show how to construct such two-valued homomorphisms. Consider first the case of finite Boolean algebras (i.e. Boolean algebras with a finite number of elements typified by the power set algebra on a set of n objects, which possesses 2^n elements).

An element a in a Boolean algebra B is called an *atom* iff $a \neq 0$ and $b \leq a \to b = 0$ or $b = a$, $\forall b \in B$. A *filter* F is a nonempty proper subset F, such that (1) $a, b \in F \to a \wedge b \in F$ and (2) $a \in F$ and $a \leq b \to b \in F$. The set of all filters in B is partially ordered by set-theoretic inclusion. An *ultrafilter* is any filter other than B itself which is maximal in this ordering. In finite Boolean algebras the ultrafilters are all generated by the atoms, i.e. they are sets of the form $\{b \in B \mid a \leq b\}$ for some atom a. Mathematicians call such ultrafilters *principal* ultrafilters. So we can express this result by saying that in a finite Boolean algebra all ultrafilters are principal.

To construct a two-valued homomorphism of a finite Boolean algebra, take any atom a and construct the associated (principal) ultrafilter F_a generated by it. Then define the map $h_a: B \to B_2$ by

$$h_a(b) = 1, \ \forall b \in F_a$$
$$= 0, \ \forall b \notin F_a$$

This map is easily seen to be a homomorphism, and it can be shown that *all* two-valued homomorphisms arise in this way. These results can be extended to arbitrary infinite Boolean algebras in the form of two theorems:

The Ultrafilter Theorem:
Every filter is contained in some ultrafilter.

The Homomorphism Theorem:
Every ultrafilter F determines a two-valued homomorphism $h_F: B \to B_2$, specified by

$$h_F(b) = 1, \; \forall b \in F$$
$$= 0, \; \forall b \notin F$$

But, in the case of infinite Boolean algebras, the ultrafilters are not in general all principal. The existence of nonprincipal ultrafilters is the complicating feature of the general case.

Representations of Boolean Algebras

Every finite Boolean algebra is isomorphic to the power set algebra of its set of atoms. This is a special case of the famous *Stone* representation theorem, which says that every Boolean algebra is isomorphic to a certain *field* of subsets of the set of all its ultrafilters. The field is not in general the full power set. (A field of subsets is just a nonempty set of subsets which is closed with respect to finite unions, intersections, and complements).

References

The first ten chapters of Halmos (1960) give a splendid introduction to those parts of naive set theory used in the present work. Alternatively Stoll (1961) may be consulted. For the theory of finite-dimensional vector spaces, the best text is Shephard (1966). More advanced but highly recommended as background for the mathematical theory of Hilbert spaces is Simmons (1963). More detail on lattice theory and the theory of Boolean algebras is given in chapter 4 of Bell and Machover (1977).

Bibliography

ARAKI, H. A. and YANASE, M. M. (1960): 'Measurement of Quantum Mechanical Operators', *Physical Review* **120**, 622–6. Reprinted in Wheeler and Zurek (1983), 707–11.

ARTHURS, E. and KELLY, J. L. (1965): 'On the Simultaneous Measurement of a Pair of Conjugate Observables', *Bell System Technical Journal* **44**, 725–9.

ASPECT, A., GRANGIER, P., and ROGER, G. (1981): 'Experimental Tests of Realistic Local Theories via Bell's Theorem', *Physical Review Letters* **47**, 460–7.

——, ——, and —— (1982): 'Experimental Realization of Einstein–Podolsky–Rosen–Bohm *Gedankenexperiment*: A New Violation of Bell's Inequalities', *Physical Review Letters* **48**, 91–4.

——, DALIBARD, J., and ROGER, G. (1982): 'Experimental Tests of Bell's Inequalities Using Time-Varying Analyzers', *Physical Review Letters* **49**, 1804–7.

BELINFANTE, F. J. (1973): *A Survey of Hidden-Variable Theories*, Oxford: Pergamon Press.

BELL, J. L. and HALLETT, M. (1982): 'Logic, Quantum Logic and Empiricism', *Philosophy of Science* **49**, 355–79.

—— and MACHOVER, M. (1977): *A Course in Mathematical Logic*, Amsterdam, New York, Oxford: North-Holland.

BELL, J. S. (1964): 'On the Einstein–Podolsky–Rosen Paradox', *Physics* **1**, 195–200. Reprinted in Wheeler and Zurek (1983), 403–8.

—— (1966): 'On the Problem of Hidden Variables in Quantum Mechanics', *Reviews of Modern Physics* **38**, 447–75. Reprinted in Wheeler and Zurek (1983), 397–402.

—— (1971): 'Introduction to the Hidden-Variable Question', in B. d'Espagnat (ed.), *Foundations of Quantum Mechanics*, Proceedings of the International School of Physics 'Enrico Fermi', Course IL. New York: Academic Press, 171–81.

BELTRAMETTI, F. G. and van FRAASSEN, B. (eds.) (1981): *Current Issues in Quantum Logic*, New York: Plenum.

BIRKHOFF, G. and von NEUMANN, J. (1936): 'The Logic of Quantum Mechanics', *Annals of Mathematics* **37**, 823–43.

BOHM, D. (1951): *Quantum Theory*, Englewood Cliffs, NJ: Prentice-Hall.

BOHR, N. (1934): *Atomic Theory and the Description of Nature*, Cambridge: Cambridge University Press.

—— (1935): 'Can Quantum Mechanical Description of Physical Reality be

Considered Complete?', *Physical Review* **48**, 696–702. Reprinted in Wheeler and Zurek (1983), 145–51.

—— (1958): *Atomic Physics and Human Knowledge*, New York: Wiley.

—— (1963): *Essays 1958–1962 on Atomic Physics and Human Knowledge*, New York: Interscience.

BOWIE, G. L. (1979): 'The Similarity Approach to Counterfactuals: Some Problems', *Noûs* **13**, 477–98.

BRODY, T. A. (1980): 'Where Does the Bell Inequality Lead?', unpublished manuscript.

—— and DE LA PEÑA-AUERBACH, L. (1979): 'Real and Imagined Nonlocalities in Quantum Mechanics', *Il Nuovo Cimento* **54B**, 455–62.

BROWN, H. R. (1986): 'The Insolubility Proof of the Quantum Measurement Problem', *Foundations of Physics* **16**, 857–70.

—— and REDHEAD, M. L. G. (1981): 'A Critique of the Disturbance Theory of Indeterminacy in Quantum Mechanics', *Foundations of Physics* **11**, 1–20.

BUB, J. (1976): 'Hidden Variables and Locality', *Foundations of Physics* **6**, 511–25.

—— (1980): 'Comment on "Locality and the Algebraic Structure of Quantum Mechanics" by William Demopoulos', in Suppes (1980), 149–53.

CLAUSER, J. F. and HORNE, M. A. (1974): 'Experimental Consequences of Objective Local Theories', *Physical Review* **D10**, 526–35.

—— and SHIMONY, A. (1978): 'Bell's Theorem: Experimental Tests and Implications', *Reports on Progress in Physics* **41**, 1881–927.

DEMOPOULOS, W. (1980): 'Locality and the Algebraic Structure of Quantum Mechanics', in Suppes (1980), 119–44.

D'ESPAGNAT, B. (1976): *Conceptual Foundations of Quantum Mechanics*, 2nd edn., Reading, Mass.: Benjamin.

—— (1984): 'Nonseparability and the Tentative Descriptions of Reality', *Physics Reports* **110**, 201–64.

DIRAC, P. A. M. (1958): *The Principles of Quantum Mechanics*, 4th edn., Oxford: Clarendon Press.

DUMMETT, M. (1976): 'Is Logic Empirical?', in H. D. Lewis (ed.), *Contemporary British Philosophy*, London: George Allen and Unwin, 45–68.

EBERHARD, P. H. (1977): 'Bell's Theorem without Hidden Variables', *Il Nuovo Cimento* **38B**, 75–80.

—— (1978): 'Bell's Theorem and the Different Concepts of Locality', *Il Nuovo Cimento* **46B**, 392–419.

EINSTEIN, A., PODOLSKY, B., and ROSEN, N. (1935): 'Can Quantum-Mechanical Description of Physical Reality be Considered Complete?', *Physical Review* **47**, 777–80. Reprinted in Wheeler and Zurek (1983), 138–41.

ELLIS, B. (1966): *Basic Concepts of Measurement*, Cambridge: Cambridge University Press.
FINE, A. (1970): 'Insolubility of the Quantum Measurement Problem', *Physical Review* **2D**, 2783–7.
—— (1974): 'On the Completeness of Quantum Theory', *Synthese* **29**, 257–89. Reprinted in Suppes (1976), 249–81.
—— (1979): 'Counting Frequencies: A Primer for Quantum Realists', *Synthese* **42**, 145–54.
—— (1981): 'Correlations and Physical Locality', in P. Asquith and R. Giere (eds.), *PSA 1980*, vol. 2, East Lansing, Michigan: Philosophy of Science Association, 535–62.
—— (1982a): 'Hidden Variables, Joint Probability, and the Bell Inequalities', *Physical Review Letters* **48**, 291–5.
—— (1982b): 'Joint Distributions, Quantum Correlations, and Commuting Observables', *Journal of Mathematical Physics* **23**, 1306–10.
—— and TELLER, P. (1977): 'Algebraic Constraints on Hidden Variables', *Foundations of Physics* **8**, 629–37.
FITCHARD, E. E. (1979): 'Proposed Experimental Test of Wave Packet Reduction and the Uncertainty Principle', *Foundations of Physics* **9**, 525–35.
FOLSE, H. J. (1985): *The Philosophy of Niels Bohr*. Amsterdam: North-Holland.
FRIEDMAN, M. and GLYMOUR, C. (1972): 'If Quanta Had Logic', *Journal of Philosophical Logic* **1**, 16–28.
—— and PUTNAM, H. (1978): 'Quantum Logic, Conditional Probability and Inference', *Dialectica* **32**, 305–15.
GASIOROWICZ, S. (1974): *Quantum Physics*, New York: Wiley.
GHIRARDI, G. C., RIMINI, A., and WEBER, T. (1980): 'A General Argument against Superluminal Transmission through the Quantum Mechanical Measurement Process', *Lettere al Nuovo Cimento* **27**, 293–8.
GLEASON, A. M. (1957): 'Measures on the Closed Subspaces of a Hilbert Space', *Journal of Mathematics and Mechanics* **6**, 885–93. Reprinted in Hooker (1975), 123–33.
GOLDBLATT, R. I. (1974): 'Semantic Analysis of Orthologic', *Journal of Philosophical Logic* **3**, 19–35.
HAACK, S. (1978): *Philosophy of Logics*, Cambridge: Cambridge University Press.
HALMOS, P. R. (1960): *Naive Set Theory*, Princeton, NJ: Van Nostrand.
HARDEGREE, G. M. (1974): 'The Conditional in Quantum Logic', *Synthese* **29**, 63–80. Reprinted in Suppes (1976), 55–72.
HARPER, W. L., STALNAKER, R., and PEARCE, G. (eds.) (1981): *Ifs: Conditionals, Belief, Decision, Chance, and Time*, Dordrecht: Reidel.
HEALEY, R. (1979): 'Quantum Realism: Naïveté is No Excuse', *Synthese* **42**, 121–44.

HEISENBERG, W. (1930): *The Physical Principles of the Quantum Theory*. Chicago: University of Chicago Press.

HELLMAN, G. (1982): 'Stochastic Einstein-Locality and the Bell Theorems', *Synthese* **53**, 461–504.

HEYWOOD, P. and REDHEAD, M. L. G. (1983): 'Nonlocality and the Kochen–Specker Paradox', *Foundations of Physics* **13**, 481–99.

HOLDSWORTH, D. G. and HOOKER, C. A. (1983): 'A Critical Survey of Quantum Logic', in *Logic in the 20th Century, Scientia Special Issue*, 127–246.

HOOKER, C. A. (ed.) (1975): *The Logico-Algebraic Approach to Quantum Mechanics*, vol. 1, Dordrecht: Reidel.

HOWARD, D. (1985): 'Einstein on Locality and Separability', *Studies in History and Philosophy of Science* **16**, 171–201.

HUMPHREYS, P. (1980): 'A Note on Demopoulos's Paper "Locality and the Algebraic Structure of Quantum Mechanics"'. In Suppes (1980), 145–7.

JAMMER, M. (1974): *The Philosophy of Quantum Mechanics*, New York: Wiley.

JARRETT, J. (1984): 'On the Physical Significance of the Locality Conditions in the Bell Arguments', *Noûs* **18**, 569–89.

JAUCH, J. M. (1968): *Foundations of Quantum Mechanics*, Reading, Mass.: Addison-Wesley.

JORDAN, T. F. (1983): 'Quantum Correlations Do Not Transmit Signals', *Physics Letters* **94A**, 264.

KITCHER, P. (1981): 'Explanatory Unification', *Philosophy of Science* **48**, 507–31.

KOCHEN, S. and SPECKER, E. (1967): 'The Problem of Hidden Variables in Quantum Mechanics', *Journal of Mathematics and Mechanics* **17**, 59–87. Reprinted in Hooker (1975), 293–328.

KRIPS, H. (1974): 'Foundations of Quantum Theory, Part I', *Foundations of Physics* **4**, 181–93.

—— (1977): 'Quantum Theory and Measures on Hilbert Space', *Journal of Mathematical Physics* **18**, 1015–21.

LANDAU, L. J. (1986): 'On the Violation of Bell's Inequality in Quantum Theory', forthcoming in *Physics Letters A*.

LEPLIN, J. (ed.) (1983): *Scientific Realism*, Berkeley: University of California Press.

LEWIS, D. (1973): *Counterfactuals*, Oxford: Blackwell.

—— (1979): 'Counterfactual Dependence and Time's Arrow', *Noûs* **13**, 455–76.

LO, T. K. and SHIMONY, A. (1981): 'Proposed Molecular Test of Local Hidden-Variable Theories', *Physical Review* **23A**, 3003–12.

MACKEY, G. (1963): *Mathematical Foundations of Quantum Mechanics*, New York: Benjamin.

MCKNIGHT, J. L. (1958): 'An Extended Latency Interpretation of Quantum Mechanical Measurement', *Philosophy of Science* **25**, 209–22.

MACZYNSKI, M. J. (1971): 'Boolean Properties of Observables in Axiomatic Quantum Mechanics', *Reports on Mathematical Physics* **2**, 135–50.

MARGENAU, H. (1954): 'Advantages and Disadvantages of Various Interpretations of the Quantum Theory', *Physics Today* **7**, 6–13.

MARSHALL, T. W., SANTOS, E., and SELLERI, F. (1983): 'Local Realism has not been Refuted by Atomic Cascade Experiments', *Physics Letters* **98A**, 5–9.

MATTHEWS, P. (1974): *Introduction to Quantum Mechanics*, 3rd edn., London: McGraw-Hill.

MELLOR, D. H. (1971): *The Matter of Chance*, Cambridge: Cambridge University Press.

MESSIAH, A. (1968): *Quantum Mechanics*, New York: Wiley.

PAGE, D. N. (1982): 'The Einstein–Podolsky–Rosen Physical Reality is Completely Described by Quantum Mechanics', *Physics Letters* **91A**, 57–60.

PARK, J. L. (1973): 'The Self-Contradictory Foundations of Formalistic Quantum Measurement Theories', *International Journal of Theoretical Physics* **8**, 211–18.

—— and MARGENAU, H. (1968): 'Simultaneous Measurability in Quantum Theory', *International Journal of Theoretical Physics* **1**, 211–83.

PERES, A. (1978): 'Unperformed Experiments Have no Results', *American Journal of Physics* **46**, 745–7.

—— and ZUREK, W. H. (1982): 'Is Quantum Theory Universally Valid?' *American Journal of Physics* **50**, 807–10.

PERRIE, W., DUNCAN, A. J., BEYER, H. J., and KLEINPOPPEN, H. (1985): 'Polarization Correlation of the Two Photons Emitted by Metastable Atomic Deuterium', *Physical Review Letters* **58**, 1790–3.

PIPKIN, F. M. (1978): 'Atomic Physics Tests of the Basic Concepts in Quantum Mechanics', *Advances in Atomic and Molecular Physics* **14**, 281–340.

PIRON, C. (1972): 'Survey of General Quantum Physics', *Foundations of Physics* **2**, 287–314. Reprinted in Hooker (1975), 513–43.

POPPER, K. R. (1934): *Logik der Forschung*, Vienna: Springer. English translation, London: Hutchinson, 1959.

—— (1957): 'The Propensity Interpretation of the Calculus of Probability, and the Quantum Theory', in S. Körner (ed.), *Observation and Interpretation*, London: Butterworths, 65–70, 88–9.

—— (1967): 'Quantum Mechanics without "The Observer"', in M. Bunge (ed.), *Quantum Theory and Reality*, Berlin: Springer, 7–44.

—— (1982a): *The Open Universe: An Argument for Indeterminism*. London: Hutchinson.

—— (1982b): *Quantum Theory and the Schism in Physics*, London: Hutchinson.

—— (1983): *Realism and the Aim of Science*, London: Hutchinson.
PRUGOVEČKI, E. (1981): *Quantum Mechanics in Hilbert Space*, 2nd edn., New York and London, Academic Press.
PUTNAM, H. (1969): 'Is Logic Empirical?', *Boston Studies in the Philosophy of Science* **5**, 216–41.
—— (1974): 'How to Think Quantum-Logically', *Synthese* **29**, 55–61. Reprinted in Suppes (1976), 47–53.
—— (1981): 'Quantum Mechanics and the Observer' *Erkenntnis* **16**, 193–219.
REDHEAD, M. L. G. (1980): 'Models in Physics', *British Journal for the Philosophy of Science* **31**, 145–63.
—— (1981): 'Experimental Tests of the Sum Rule', *Philosophy of Science* **48**, 50–64.
—— (1983): 'Relativity, Causality and the Einstein–Podolsky–Rosen Paradox: Nonlocality and Peaceful Coexistence', in R. Swinburne (ed.)
—— (1984): 'Undressing Baby Bell', forthcoming in R. Bhaskar (ed.) *Realism and Human Being*, Oxford: Blackwell.
—— (1986): 'Relativity and Quantum Mechanics—Conflict or Peaceful Coexistence', *Annals of the New York Academy of Sciences*, **480**, 14–20.
—— (1987): 'Whither Complementarity?' in N. Rescher (ed.) *Scientific Inquiry in Philosophical Perspective*, Lanham: University Press of America, 169–82.
REICHENBACH, H. (1928): *Philosophie der Raum–Zeit–Lehre*, Berlin: Walter der Gruyter. English translation, New York: Dover, 1958.
—— (1956): *The Direction of Time*, Berkeley and Los Angeles: University of California Press.
ROBERTSON, H. P. (1929): 'The Uncertainty Principle', *Physical Review* **34**, 163–4. Reprinted in Wheeler and Zurek (1983), 127–8.
SALMON, W. (1984): *Scientific Explanation and the Causal Structure of the World*, Princeton: Princeton University Press.
SCHEIBE, E. (1973): *The Logical Analysis of Quantum Mechanics*, Oxford: Pergamon.
SCHIFF, L. (1968): *Quantum Mechanics*, 3rd edn., London: McGraw-Hill.
SELLERI, F. and TAROZZI, G. (1980): 'Is Clauser and Horne's Factorizability a Necessary Requirement for a Probabilistic Local Theory?', *Lettere al Nuovo Cimento* **29**, 533–6.
SHE, C. Y. and HEFFNER, H. (1966): 'Simultaneous Measurement of Noncommuting Observables', *Physical Review* **152**, 1103–10.
SHEPHARD, G. C. (1966): *Vector Spaces of Finite Dimension*, Edinburgh and London: Oliver & Boyd.
SHIMONY, A. (1978): 'Metaphysical Problems in the Foundations of Quantum Mechanics', *International Philosophical Quarterly* **18**, 3–17.
—— (1981): 'Critique of the Papers of Fine and Suppes', in P. Asquith and R.

Giere (eds.), *PSA 1980* vol. 2, East Lansing, Michigan: Philosophy of Science Association, 572–80.

—— (1984): 'Controllable and Uncontrollable Non-Locality' in S. Kamefuchi *et al.* (eds.) *Proceedings of the International Symposium: Foundations of Quantum Mechanics in the Light of New Technology,* Tokyo: Physical Society of Japan, 225–30.

SIMMONS, G. F. (1963): *Introduction to Topology and Modern Analysis,* New York: McGraw-Hill.

SLOTE, M. A. (1978): 'Time in Counterfactuals', *The Philosophical Review* **87,** 3–17.

STAPP, H. (1971): 'S-Matrix Interpretation of Quantum Theory', *Physical Review* **D3,** 1303–20.

STOLL, R. R. (1961): *Sets, Logic and Axiomatic Theories,* San Francisco: Freeman.

SUPPES, P. (ed.) (1976): *Logic and Probability in Quantum Mechanics,* Dordrecht: Reidel.

—— (ed.) (1980): *Studies in the Foundations of Quantum Mechanics,* East Lansing, Michigan: PSA.

—— and ZANOTTI, M. (1976): 'On the Determinism of Hidden Variable Theories with Strict Correlation and Conditional Statistical Independence of Observables', in Suppes (1976), 445–55.

SVETLICHNY, G., REDHEAD, M. L. G., BROWN, H. R. and BUTTERFIELD, J. (1988): 'Do the Bell Inequalities Require the Existence of Joint Probabilities?', *Philosophy of Science,* **55,** 387–401.

THOMASON, R. and GUPTA, A. (1980): 'A Theory of Conditionals in the Context of Branching Time', *The Philosophical Review* **89,** 65–90. Reprinted in Harper *et al.* (1981), 299–322.

VAN FRAASSEN, B. (1972): 'A Formal Approach to the Philosophy of Science', in R. Colodny (ed.), *Paradigms and Paradoxes,* Pittsburgh: University of Pittsburgh Press, 303–66.

—— (1973): 'Semantic Analysis of Quantum Logic', in C. A. Hooker (ed.), *Contemporary Research in the Foundations and Philosophy of Quantum Theory,* Dordrecht: Reidel, 80–113.

—— (1974): 'The Labyrinth of Quantum Logic', *Boston Studies in the Philosophy of Science* **13,** 224–54. Reprinted in Hooker (1975), 577–607.

—— (1979): 'Hidden Variables and the Modal Interpretation of Quantum Theory', *Synthese* **42,** 155–65.

VARADARAJAN, V. S. (1968): *Geometry of Quantum Theory,* 2 vols., Princeton, NJ: Van Nostrand.

VON NEUMANN, J. (1932): *Mathematische Grundlagen der Quantenmechanik,* Berlin: Springer. English translation, Princeton, NJ: Princeton University Press, 1955.

WHEELER, J. A. and ZUREK, W. H. (eds.) (1983): *Quantum Theory and Measurement*, Princeton, NJ: Princeton University Press.
WIGNER, E. P. (1952): 'Die Messung quantenmechanischer Operatoren', *Zeitschrift für Physik* **133**, 101–8.
—— (1970): 'On Hidden Variables and Quantum Mechanical Probabilities', *American Journal of Physics* **33**, 1005–9.
YANASE, M. M. (1961): Optimal Measuring Apparatus', *Physical Review* **123**, 666–8. Reprinted in Wheeler and Zurek (1983), 712–14.

Supplementary Bibliography

BELL, J. S. (1987): *Speakable and Unspeakable in Quantum Mechanics*, Cambridge: Cambridge University Press.
CLIFTON, R. K., BUTTERFIELD, J. and REDHEAD, M. L. G. (1988): 'Nonlocal Influences and Possible Worlds—A Stapp in the Wrong Direction', forthcoming in *The British Journal for the Philosophy of Science*.
CLIFTON, R. K. and REDHEAD, M. L. G. (1988): 'The Compatibility of CP Violating Systems with Statistical Locality', *Physics Letters A*, **126**, 295–9.
GIBBINS, P. (1987): *Particles and Paradoxes: The Limits of Quantum Logic*, Cambridge: Cambridge University Press.
HONNER, J. (1987): *The Description of Nature: Niels Bohr and the Philosophy of Quantum Physics*, Oxford: The Clarendon Press.
KRIPS, H. (1987): *The Metaphysics of Quantum Theory*, Oxford: The Clarendon Press.
MURDOCH, D. (1987): *Niels Bohr's Philosophy of Physics*, Cambridge: Cambridge University Press.
REDHEAD, M. L. G. (1988): 'Nonfactorizability, Stochastic Causality and Passion-at-a-Distance', forthcoming in J. Cushing and E. McMullin (eds): *Philosophical Consequences in Quantum Theory*, Notre Dame: Notre Dame University Press.
STAPP, H. (1988): 'Quantum Nonlocality and the Description of Nature', forthcoming in J. Cushing and E. McMullin (eds): *Philosophical Consequences in Quantum Theory*, Notre Dame: Notre Dame University Press.

Index

Note: **named Principles are highlighted by enboldened page numbers**; 'QM' stands for 'quantum mechanics' in this index (as it has throughout the book).

action-at-a-distance 107, 168, 169
adjoint (of linear operator), definition 173
Admissibility criteria (for truth valuations)
  A$_1$ admissibility criterion 160
  A$_2$ admissibility criterion 165
alternative logics, approaches to 154
analogue model, necessity of 45
angular momentum 32–9
  of composite system 40–2
  literature references on 43
Aristotelian physics 48
Aspect experiment 108, 109, 110–13
  light cone structure for 111
  optical switch used in 112
  schematic representation 110
atom (in Boolean algebra) 177
axioms 155

Banach algebra 26–7
Bell experiment 83–4
  Aspect version 110–13
  γ-ray polarization correlation versions, 107, 108–9
  low-energy proton–proton scattering versions 107, 108
  visible photon correlation versions 107, 108, 109–13
Bell inequality 82–90
  alternative forms of 96–8
  assumptions made 89–90
    determinism assumption 89, 90, 168–9
    randomness assumption 89, 90
  assumptions not made 90
  experimental tests of 107–113
  indeterministic proof of 98–101
  Matching Condition for **88, 91**
  Stapp–Eberhard approach 90–6
  violated in QM 85, 87
Bell locality 75
  violation of 110

bibliography 179–85
Bohr's views on QM 1, 50–1
  literature references 69
Boolean algebras
  definition 176
  homomorphisms of 177–8
  representations of 178
Boolean lattice
  definitions 23, 176
  replacement by projection lattices of Hilbert space 23, 160
Boolean sublattices 23, 25
bras, definition 6

Cartesian product (of sets), definition 170
characteristic algebra 156, 167
classical physics, logic of 157–60
classical propositional calculus (CPC)
  semantics produced for 155–6
  syntactic notions for 155
classical propositional logic 155–7
Clebsch–Gordon coefficients 40
Comeasurable Value Rule (CVR) 143
  derivation from Value Rule and FUNC* 144
  incompatibility with locality 146–50
commutation relations 9, 32
commuting observables
  quantum-mechanical joint probability distribution 18
complement (of set), definition 170
complementarity
  Bohr's view 51, 69
  definition 50
  interpretation of QM 49–51
complemented lattice, definition 176
completeness conditions
  EPR Argument 71
  Hilbert space 175
  stochastic hidden-variable theories 99
Completeness Metatheorem 155
  symbolic expression 156

Index

connected manifold, definition 29
conterfactual conditional
  symbol used 93
  truth conditions for 93–4
contextuality
  environmental 140
  connection with nonlocality 139–42
  ontological 135
continuity, topological definition 29
Copenhagen complementarity interpretation 49–51
Correspondence Rule (Corr) 133
  rejection to escape Kocher–Specker paradox 134–5
counterfactuals 88, 90–6, 151
  literature references 93, 117

de Morgan laws 163
Deduction Metatheorem 154, 164
definite, meaning of term 95
determinism, assumption in Bell inequality 89, 90, 168–9
Dirac delta function 10
Dirac formulation 6–12
  quantization algorithm 7
  statistical algorithm 8–9
Dirac picture 12, 116
Dirac vector notation 6, 173
dispersion-free states, impossibility in QM 62, 70
distributive lattice, definition 176
Disturbance Theory of uncertainty relations 67, 70
domain, definition 171

eigenvalue, definition 173–4
eigenvalue spectrum 8
Eigenvector Rule 73, 120
Einstein Dilemma 76, 82
Einstein locality 75
  violation of 110, 111, 112, 113
Einstein–Podolsky–Rosen (EPR) Argument 71–81
  Bohr's response 1, 77, 81
  Completeness Condition (P) 71
  Reality Condition (R) 72
  schematic illustration 76
  spin version 73–4, 78–81
  summary equation 75–6
Einstein's Reality Criterion (R') 78
elementary propositions 158, 164
empty set, definition 170
environmental contextuality 140

Environmental Locality (ELOC) 141
  violation of 151, 168
Euclidean 3-space 28
explanation, nature of 45, 69
Extended Value Rule (EVR) 142–3

factorizability, stochastic hidden-variable theories 100
Faithful Measurement Principle (FM) 89, 99, 119–20
formalism of QM 5–43
Functional Composition Principle (FUNC) 121
  justification of 131–8
  relationship to STAT FUNC 132
FUNC* (restricted version of FUNC) 137
fuzzy-valued observables 48

γ-ray microscope example 67
γ-ray polarization correlation experiments 107, 108–9
Gleason's theorem 27–30, 128
Gödel result 71

Hasse diagrams, 24, 25
  definition 23
Heisenberg picture 12
hidden-variable interpretation 1, 45–8
Hilbert space
  Boolean sublattice of, 25
  colouring of hypersphere surface in 123
  definition 175–6
  projection lattice of 22, 23, 24, 26, 27, 160
  three-dimensional 25, 28
Homomorphism Theorem 157, 178
homomorphisms, of Boolean algebras 177–8

ideal measurement 52, 59, 69
idealism 49
idempotent operator 174
ignorance interpretation of mixtures 53, 69
incompleteness argument of Einstein, Podolsky and Rosen 71–81
indeterminism 90–6, 98
interpretation, meaning of word 44, 69
interpretations (of QM) 44–69
  view A: hidden variables 45–8, 76–7, 89, 117, 119, 133, 134
  view B: propensities/potentialities 48–9, 76–7, 91, 117, 168

Index

view C: complementarity **49–51**, 76–7, 91, 117, 168
intersection (of set), definition 170
inverse function, definition 171
inverse image, definition 171

kets, definition 6
Kochen–Specker diagrams
  definition of 126
  inconsistent 129, 130
Kochen–Specker Paradox 119–38
  definition 121
  demonstration of 121–31
  and nonlocality 139–51
  ways of escaping from 133–6

latent-quantity concept 48, 69
lattice, definitions 22, 176
lattice theory 176–8
Lo-Shimony experiment 109, 118
locality assumptions
  stochastic hidden-variable theories 100
Locality Principles
  Environmental Locality (ELOC) 141
  L formulation 75
  LOC_1 formulation 77, 117
  LOC_2 formulation 77, 117
  LOC_3 formulation 82, 117
  LOC_4 formulation 91, 117
  LOC_5 formulation 113, 117
  Ontological Locality (OLOC) 139–40
locally maximal observables 139
  Maczynski's theorem extended to 140, 151–2
logic, revisability of 153–4
logical connectives
  lattice operation equivalences 160
  set-theoretic operator equivalences 158
  symbols for 155
logical equivalence, symbol for 161
logico-algebraic approach 22–7
  contrasted with algebraic approach 26–7
  definition 26
  literature reference for 43
low-energy proton–proton scattering experiment 107, 108, 109
lowering operator 33

Maczynski's theorem 134, 138
  extension to locally maximal observables 140, 151–2

manifold, definition 29
many-particle systems, quantum mechanics of 30–2
map, definition 171
Matching Condition (for proving Bell inequality) 88, 91–2
material implication
  operation of 164
  symbol for 162
mathematical appendix 170–8
matrix mechanics 10
matrix representation 173
maximal observables
  FUNC* consistent with algebraic structure of 138
  locally maximal 139
measurement
  classical physics view 51, 52, 69
  quantum theory of 51–9, 69
Metatheorem (M) 157
minimal instrumentalist interpretation of QM 44
minimal quantum logic (MQL) 161
mixed state 15
mysterious nature of QM 50

nonlocality
  and Bell inequality, 82–118
  and Kochen–Specker paradox 139–52
nonseparability 106, 139, 151, 168, 169

observables
  definition 5
  functions of 16–22
  value assignments for 119
ontological contextuality 135
ontological determinism 90, 113
Ontological Locality (OLOC) 139–49
  violation of 150–1
operators, notation for 7
optical-photon correlation experiments 107, 108, 109–13
optical switch 110, 112
orthocomplement, conditions for 176–7

passion-at-a-distance 106–7, 118
Pauli spin operators 34
peaceful coexistence (between non-locality and special relativity) 168, 169
picturability 45
Planck's constant 9
Poisson bracket, definition 9

poset (partially ordered set), definition 176
potentiality interpretation of QM 48–9
power set, definition 170
power set algebra/lattice 176
  Hasse diagrams for 24, 25
pragmatic determinism 90
Principle of Local Counterfactual Definiteness (PLCD) 92
  meaning of 'definite' 95
probability measures 27
Product Rule 123, 138
projection lattices 22, 23, 24, 160
projection operators
  definitions 12–13, 15, 174
  as functions of observables 17
  orthogonality properties 16
  role played by 22
Projection Postulate 58, 59, 69
propensities 48–9, 69
proper subset, definition 170
pure state 15
Putnam states
  definition 164
  and realism 164–7

quantization algorithm 5, 120
  Dirac formulation 7
  von Neumann formulation 12
quantum logic (QL)
  contrasted with classical logic 161
  definitions 26, 153, 160
quantum mechanics (QM)
  angular momentum in 32–9
  formalism of 5–43
  interpretation of 44–70
  many-particle systems 30–2, 40–2
quantum propositional logic 160–4

Radon–Nikodym theorem 30
raising operator 33
randomness, assumption in Bell inequality 89, 90
range, definition 171
rays
  definition 7
  projection operators, relationship to 21–2, 23
realism
  and Putnam states 164–7
  and quantum logic 153–67
realist interpretations of QM 47, 48, 49
Reality Condition (EPR Argument) 72

Reality Criterion, Einstein's formulation (R') 78
Reality Principle (for proof of FUNC) 133
rejection to escape Kochen–Specker paradox 135–6
Riesz–Fischer theorem 176
robustness condition, stochastic causality, 102–6
rules of inference 155

scalars, definition 171
Schrödinger picture 12
self-adjoint operator
  definition 173
  function of 174
semantic approach (to alternative logics) 154
separability 107, 168, 169
  Hilbert space 175
  see also **Ontological Locality**
set theory 170–1
singlet state
  rotational invariance of 40–1
  two spin–1 particles 42
  two spin–$\frac{1}{2}$ particles 40–2, 73–4
$SO(3)$ 104
special relativity (SR), and nonlocality 75, 168, 169
spectral theorem 13, 174
  alternative proofs 13–14
Spectrum Rule 120
spin, definition 32
spin Hamiltonian 38, 131, 148
spin quantum numbers 40
Stapp–Eberhard approach (to Bell inequality) 90–6
state of system
  Putnam state 164
  QM state 5, 164
state-preparation aspect of measurement 52, 58, 59
statistical algorithm 5
  Dirac formulation 8–9
  von Neumann formulation 12, 14–16
Statistical Functional Composition Principle (STAT FUNC) 18
  relationship to FUNC, 131–2
statistical nonlocality 113–16
statistical operator 15
  measurement theory 53–60
stochastic causality, robustness condition for 102–6

Index

stochastic hidden-variable theories 98–107
  completeness assumption 99
  factorizability 100
  literature references 117–18
  locality assumptions 100, 107
Stone representation theorem 178
string, definition 155
$SU(2)$ 104
subset, definition 170
Sum Rule 121, 138
syntactic approach (to alternative logics) 154

tensed counterfactuals 93–4
tensor products
  definitions 174–5
  Dirac notation 31
  of linear operators 30–1, 175
  of vector spaces 30, 175
  of vectors 30, 174–5
theoremhood, symbol for 156
theories, nature of 44
time evolutions in the theory of measurement 55–6
time-dependent Schrödinger equation 12
time-independent Schrödinger equation 11
time-of-flight example 64–7
Trace measure 30
Two-Colour Theorem 124
  Belinfante's proof 125
  falsity in two dimensions 124–5

$U(1)$ 104
$U(2)$ 104

ultrafilter, definition 177
Ultrafilter Theorem 157, 178
uncertainty relations 59–69
  Disturbance Theory 67, 70
  literature references on 70
union (of set), definition 170

validity, symbol for 156
value assignments, QM observables 119
Value Rule (VR) 120, 142, 150
van Fraassen split-observable approach 134–7, 139
vector space
  definition 171
  inner product on 172
  linear functional on 172
  linear operator on 172
vector space theory 171–6
vectors, Dirac notation 6, 173
view A (hidden-variable interpretation of QM) 45–8, 76–7, 89, 117, 119, 133, 134
  rejection to escape Kochen–Specker paradox 133–4
view B (potentiality interpretation of QM) 48–9, 76–7, 91, 117, 168
view C (complementarity interpretation of QM) 49–51, 76–7, 91, 117, 168
von Neumann formulation of QM 12–16
  quantization algorithm 12
  statistical algorithm 12, 14–16
von Neumann proof of the impossibility of hidden variables 1

wave function, definition 10
well-formed formula (wff), definition 155